The Oxford Hill Press

MPS
MANAGED PRINT SERVICES

The Oxford Hill Press

MPS

MANAGED PRINT SERVICES
INSIGHT AND BEST PRACTICES FOR BUYERS AND SELLERS AS YOU NAVIGATE THE COMPLEXITIES OF OFFICE OUTPUT OPTIMIZATION INITIATIVES

TAB EDWARDS

A TAB REPORT BOOK
PUBLISHED BY THE OXFORD HILL PRESS

Philadelphia • New York • Chicago • Toronto

OXFORD HILL PRESS
A Division of The Oxford Hill Consulting Group

Copyright © 2010 by Tab Edwards. All rights reserved.

Printed in the United States of America. Except as permitted under the United States Copyright Act of 1976, no part of this publication may be reproduced or distributed in any form or by any means, or stored in a data base or retrieval system, without the prior written permission of the publisher.

ISBN 978-0-9700891-7-5

This publication is designed to provide authoritative information in regards to the subject matter covered. It is sold with the understanding that the publisher is not engaged in rendering legal, accounting, or other professional services. If legal advice or other expert assistance is required, the services of a competent professional person should be sought.

—From a declaration of principles jointly adopted by a committee of the American Bar Association and a committee of publishers.

Oxford Hill Press books are available at special quantity discounts to use as premiums and promotions, or for use in corporate training programs. For more information, please call 413.502.3751, or contact Monise Gersh at monise.gersh@oxfordhill.com.

Designed by Joshua Black of Blackeyesoup.com
Philadelphia, PA.

1 3 5 7 9 10 8 6 4 2

To my mother,
Angelean Edwards

The Oxford Hill Press

CONTENTS

Introduction   13

ONE
The Scope of this Book:
The General Office   18

TWO
Managed Print Services Defined   21

THREE
The Shift Toward
Managed Print Services   27

FOUR
The Elements of a
Managed Print Services Solution   39

FIVE
The Managed Print Services
Process   49

SIX
Total Cost of Ownership vs. Cost-Per-Page   58

SEVEN
Drawing Distinctions   67

EIGHT
The 7Cs of Imaging & Output   75

NINE
Managed Print Services as a
Green Initiative   140

TEN
MPS Buyers: How to Get Started
(on the Road to MPS)   150

ELEVEN
MPS Sellers: How to Get Started
(on the Road to MPS)   170

TWELVE
Best Practices for Managing
and Controlling
Personal Printers in the General Office   182

THIRTEEN
Insider Trading   191

FOURTEEN
The Question   200

About the Author   203

Index   205

MPS
MANAGED PRINT SERVICES

The Oxford Hill Press

INTRODUCTION

Flashback to the year 2005. I was in the embryonic stages of working on a Managed Print Services engagement with a top-5 global Pharmaceutical manufacturer (that I will call "Giant Pharma") that was interested in determining how they could reduce their cost of output while streamlining the device fleet which had ballooned to more than 42 different models of printers and 45 different models of copiers & MFPs from eleven different hardware manufacturers. The company's project leader (whom I will call "Bob") assumed that since the environment was "way out of control" there *must* be a significant opportunity to reduce the company's cost of printing, copying and faxing (Imaging & Output).

It's worth pointing out that Bob and I first began discussing and investigating the opportunity for improvement in this company three years earlier in 2002 when the concept of Managed Print Services or *any* type of bundled print-related solution was in the nascent state. Bob believed that the company — which was looking to reduce operating costs — could surely do that by focusing on Imaging & Output. However, he was having a difficult time getting upper-management's attention about investing in the pursuit of MPS. To these managers, printers and copiers were not "critical or expensive" (or so they assumed) and therefore, it wasn't worth their time to consider.

Frustrated, Bob pleaded his case to an internal management ally who was able to get approval to study one 3-story, out-of-the-way building in New Jersey. Bob called me on the phone and excitedly said, "We can get started!"

I conducted a thorough assessment of the building and presented the findings and my improvement-recommendation to Bob and his ally manager. In short, my findings showed that Giant Pharma could reduce its hard-dollar operating costs in this one building by 42% and that if we extrapolated the findings to represent the other U.S. locations, Giant Pharma could save approximately $4.2M in the general office alone; this was enough to get the attention of the previously-indifferent executive management team. You would think that with those kinds of numbers, the executives would be giddy with excitement over the prospect of reducing their operating costs by $4.2M. Wrong. The management was "intrigued," but not sold. Yet.

They had lots of questions: How was the assessment conducted and why should we believe the results are an accurate reflection of our current state performance? And what exactly is "Managed Print Services" *anyway* and how do we know it will deliver the expected results? Why should we take away personal printers from our high-priced consultants and make them "waste time" walking to a shared printer to retrieve their output? What about competitive offerings? What happens if this MPS thing goes wrong, what are the risks? And who the hell is Tab Edwards and why should we believe *him*?

Of all of the assessments that I had conducted to that point, none required the acme of diligence and perfection as this project. Not only was Bob's reputation on the line for sticking his neck out and almost demanding that he be heard on this matter, but also *my* reputation as a credible MPS consultant was hanging in the balance. But we persevered. Based on the potential cost savings of the project Bob was given permission to pilot my proposed MPS solution at the Information Technology (I/T) facility. During the course of the pilot period we had to respond to the Chief Information Officer's objections and we nailed every one of them. How? Primarily by validating my assessment methodology, my assessment cost-calculation tool and by Building the Business Case for why Giant Pharma should move forward with the project. This Business case addressed the issues of risk, implementation time, cost-savings validation, efficiency improvement, user resistance, Return on Investment, the expected changes with the MPS solution in place and — you guessed it — explaining why it would actually be cheaper and faster for those "high-priced consultants" to walk 15-feet to retrieve their printed output from a shared printer.

In the end, we won the multi-million dollar business — globally — and Bob ...? Let's just say that someone else at Giant Pharma took credit for the MPS project and its multi-million dollar cost savings, but Bob, eventually, got a promotion out of the deal. However, Bob did get a personal win out of this process, too. Giant Pharma's initial interest was only in replacing their analog copiers (yes, they had tons of them), but Bob took great pride in the fact that, working together, we were able to convince the executive management team that they should consider revamping *all* of it; which they ultimately did.

This engagement required near-perfection and it was not for the faint-of-heart. It required EVERY bit of MPS know-how that I could muster just to stay in the game. It is rare that any MPS seller will face this level of high-stakes scrutiny in their career and, because of that, most sellers will not benefit from the value of this type of experience. That is the primary purpose of this book; to share best-practices from my experiences in these types of large-scale, scary, MPS engagements so that you, the seller, will at least be familiar with how to approach a situation such as this as well as smaller MPS deals should the need arise. And for the Buyer, I also share my experiences from your perspective, having worked *for* MPS-buying clients and *with* MPS-buying clients as an advisor and consultant.

I am sharing the value of my and other top consultants' experiences and offering my opinion on this topic of Imaging & Output/MPS and how companies — both buyers and sellers — can consider the many complexities of Managed Print Services in hopes of having a positive experience engaging with the solution. Is everything I write herein *gospel*? No. But what I hope to achieve with this work is to make life a little easier for professionals as you invest in Managed Print Services and to share with you some valuable insight that will allow you to avoid some pot-holes that could prove to be costly.

Two Perspectives

You will notice that I offer insight into Managed Print Services from the perspectives of both the companies that *sell* Managed Print Services solutions and the companies that *buy* MPS solutions from the sellers. I believe that if one is going to write a book on a given topic of interest to multiple stakeholders, the book should provide information that is valuable to as many of those stakeholders as is possible. My editor implied that writing this book and giving insight into best practices for both *sellers* of MPS and the companies they are selling the solution *to* is like "opening the kimono" to each side and somehow sharing each other's trade secrets. My response was: *And ...?*

Well, in the end, I won out, so this book offers insight and best-practices for both the buyers and sellers of Managed Print Services. From the buyers' perspective, you will gain insight into the sellers' motivation for offering and selling MPS solutions. And from the sellers' perspective, you will gain insight into what is important to customers as they investigate and evaluate the possible purchase and implementation of a MPS solution. With both parties having an understanding of the other's motivation, I believe that it will result in a better working relationship between the two, a clear understanding of what each other's motivation is and — if we're lucky — that insight will serve as a platform for ensuring that the sellers are focused on what's important to the buyers (the buyer's business objectives), while finding a way to support the buyer with Imaging & Output-related services (such as MPS) that they can sell. Never let it be said that I am not doing my part to bring *harmony* to the buyer-seller relationship!

This book is neither intended to be a panacea nor provide the answers to all the world's mysteries surrounding Imaging & Output and MPS. I do believe, however, that the information contained herein will serve you well in your endeavor.

ONE

The Scope of This Book: The General Office

The scope of this book is Imaging & Output in the general business office and the production of printed output therein by the workers. "Imaging & Output" in this context relates to the production of the printed page and/or the capture of the printed page to be reproduced as a digital image (for creating copies or for transmitting electronically).

When talking about the different types of printing that takes place in a business, the scope can be as simple as an Administrative Assistant printing a 1-page meeting agenda to an Advertising agency outsourcing collateral production to a third-party printer on their client's behalf, to a Pharmaceutical manufacturer printing a multi-thousand page New Drug Application to propose that the FDA approve a new pharmaceutical for sale and marketing.

As you can imagine, there are different types of requirements, printing devices, processes, infrastructures and other considerations required to satisfy the disparate printing needs of these different types of printed-output users. And when it comes to a discussion of "Managed Print Services" it should come as no surprise that there is often confusion about the boundaries of Managed Print Services — assuming there are such boundaries. Is Managed Print Services an offering for all printed output throughout a company? Does it include the printed output that is created through a Marketing or Advertising agency on behalf of a company? Does it include the printed output that a company sends off-site to be printed at a third-party print service provider? How about document management and automating paper-based processes converting them to digital? They all involve printed output and "Imaging & Output" devices at some point in the process, so shouldn't that make them fair-game for a Managed Print Services solution?

The answer — as I am sure you have guessed by now is — it depends. Or how about this: Yes and No. Or maybe this: It could, but not typically (this, by the way, is the answer that I typically give). The reality is that Managed Print Services can extend to any boundaries limited only by the solution provider's ability to deliver a comprehensive set of hardware, software, supplies & consumables, services, support and infrastructure management to manage and support the company. So, if a company like, say Hewlett-Packard has the ability to provide print management services across a company's entire print-related operation, then yes, Hewlett-Packard's definition of Managed Print Services would be all-encompassing to include general office printing, production printing and outsourced printing. If, on the other hand, Joe's Imaging & Output Company can only support a company's general office environment, then in Joe's case, the definition and scope of Managed Print Services would only extend to the general office.

The General Office

Throughout this book I will refer to Managed Print Services in the context of the General Office, where I would estimate more than 90% of Managed Print Services engagements are being fulfilled (in terms of the number of overall engagements). So what is the "General Office"? As I use the term herein, I am referring to the General Office as a facility within a company or business where the (normally) white-collar employees complete their work tasks. Or it can simply be described as the general office work environment where employees have local, easy access to the use of printers, convenience (self-service) MFPs & copiers and fax machines. The functional business areas that work out of a General Office include administrative workers, finance, accounting, marketing, sales, I/T, purchasing, real estate and other "professional" job functions. Though this is the primary scope of Managed Print Services for this book, I want to point out that other functional areas of a business are also within the scope of the "General Office" including lab workers and Scientists, academicians, healthcare professionals and even — though to a smaller degree — warehouse employees who create printed output in the course of performing their jobs.

Many of you are familiar with my use of the term "The Document-related Information Supply Chain" or DISC, to describe work-related processes that involve the creation of printed output and the involvement of paper at some point along the process. The General Office is an element of the Document-related Information Supply Chain, however, the General Office can have its own DISC and the DISC can be (and almost always is) inclusive of the General Office environment.

TWO

Managed Print Services Defined

What is Managed Print Services? Before we can move forward with a discussion of Managed Print Services we must first understand exactly what Managed Print Services is. As I go across the globe working with companies on Document-related Information Supply Chain (DISC) issues and specifically, general office Imaging & Output related issues I am not surprised when I hear someone describe Managed Print Services as something it is not. So what is Managed Print Services?

Managed Print Services — or MPS as it is most commonly referred to — is a comprehensive, bundled solution that provides convenient, reliable output to a company's users. A true Managed Print Services solution offering requires the service provider to take primary responsibility (which typically involves fleet ownership) for meeting the customer's office printing needs through a combination of hardware, supplies & consumables, service & support and overall fleet management. I acknowledge that this definition is somewhat broad and could leave the door open to many offerings being considered "Managed Print Services." But, of course, I would be remiss if I didn't provide you with a technical definition of what I believe to be Managed Print Services.

But I intend for this book to be practical and not theoretical, so I will take the definition provided in the paragraph above and break it down into its practical elements in order to provide you with a practical, working definition of what I consider to be a true Managed Print Services offering.

If we break-down the elements of the definition I provided for Managed Print Services, they would include hardware, supplies & consumables, software, service & support and management (a comprehensive list of the elements that can comprise a Managed Print Services solution offering is provided in Chapter 4: *The Elements of a Managed Print Services Solution*):

Hardware

The devices that create the printed page, duplicate the page and/or transmit the page. This includes single-function printers (including label, impact and thermal printers), analog copiers (yeah, there are still some of those around), analog fax machines, multi-function peripherals or "MFPs" (also referred to as multi-functional devices or "MFDs". When I use the term in this book I will refer to them as "MFPs.") and print server appliances. What is a "print server appliance"? A Print Server Appliance is a server appliance designed to easily manage an organization's network printing. It can offload the print traffic from the main backbone of the network and make print queue creation and driver management simple. Basically, it is an inexpensive option to the traditional PC-based print server.

Supplies & Consumables

This includes the things that the printers and other hardware need to produce printed output, including ink, toner, ribbons, paper, replacement parts for the printer/MFP (fusers, drum units, rollers, etc.), preventative maintenance kits and the like.

Software

Includes software tools for managing the device fleet such as Pharos Blueprint, Staples Technology Solutions' Printuition, Xerox's CentreWare Web, Hewlett-Packard's Web Jetadmin and other similar tools. Software can also include applications for streamlining Faxing, scan-to-email, device driver installation, queue management, solution & job management, charge-back-bill-back and the like.

Service & Support

This primarily includes options for ensuring that the device fleet is available for use with high service levels, including post-warranty device agreements, service contracts, break-fix support, service level guarantees and help desk support options. On the service side it can include any professional services that can make the experience a positive one, such as Project Management, on-site Administration, device disposition services.

Management

MPS Management includes services that take the responsibility for managing the day-to-day operation of the MPS contract (and hence the device fleet, supplies & consumables replenishment and service level guarantees) away from the customer. Depending on the scope of the MPS agreement, "management" could be as simple as managing the terms of the contract or tracking the hardware fleet and its usage, through a complete outsource of the Imaging & Printing infrastructure — including reporting.

The Benefits of Managed Print Services

Benefits To the Buyer

—*Operational Cost Savings*

The stated hard-dollar operational cost-savings delivered by MPS solutions are between 9% and 33% with an average of 23%. The range of cost savings that I have been able to show customers ranged anywhere between 2% on the very low end and in excess of 45% on the high end, with an average in the 34% range. MPS cost savings are derived from the following areas:

- Reduced I/T support costs for Imaging & Output user-related issues (Print-related help desk calls can be reduced by 50%)
- Reduced costs for supplies
- Reduced costs for print/copy/fax/scan repairs
- Reduced cost to install and upgrade hardcopy devices
- Reduced hardcopy device equipment costs. (An important point to note is that right-sizing alone won't maximize or allow a company to maintain, savings!).

—Indirect, "Soft" Cost Savings (or "productivity" cost savings)

Productivity cost savings (for those of you who care about such things) benefits are usually the greatest and consist of:

- Print-related help desk calls decreased, with more first-call resolutions
- Device availability increased
- Document workflows improved with the use of MFPs

—Improved User Productivity

—Improved fleet reliability and device uptimes

—User Satisfaction can increase in excess of 70% (based on my MPS user studies)

—Knowledge

Companies gain an understanding of not only the infrastructure specifics of their environment (the number of devices, the number of pages produced, where there is excess capacity and their Total Cost of Ownership for Imaging & Output), but they also gain an understanding of methodologies for making improvements, reducing costs and maintaining an optimized environment.

—Freedom!

By allowing the MPS solution provider to manage your Imaging & Output environment for you, you can be freed-up to focus your efforts on other critical activities with the comfort of knowing that your Imaging & Output environment will deliver some desirable level of service and device uptimes to your company's users.

—Being more "Green" and environmentally responsible

(see Chapter 8: *Managed Print Services as a Green Initiative*)

Benefits to the Seller

—*Greater Profit Margins*

Managed Print Services deals are a higher-margin revenue opportunity than selling printers and toner transactionally.

—*Stronger Customer Relationships*

You gain the ability to forge long-term strategic relationships with your MPS customers for the duration of the MPS agreement (typically 3 to 5-years).

—*Constant Revenue Stream*

MPS provides the solution provider with an ongoing revenue stream for the duration of the MPS agreement. This helps with planning and forecasting.

—*Competitive Strength*

Ability to win more business from non-MPS competitors. If the solution provider is providing exceptional service to its MPS customer, that customer will be more inclined to add additional areas of the company to the solution provider's MPS contract, including competitor's business.

—*Experience!*

The more experience the solution provider gets with delivering and managing MPS successful engagements, the better the solution provider will become at selling MPS, offering MPS and delivering a great MPS experience to its customers, leading to more customer references and more engagements.

THREE

The Shift Toward Managed Print Services

Gartner predicts that by the year 2012, over 70% of businesses with more than 250 employees will adopt a managed print services program. And estimates from Water suggest that as much as 24% of all Imaging & Output hardware that is purchased is done under some type of utility model like MPS. So why has there been such interest and growth in MPS over the past several years? The growth in interest of Managed Print Services is the result of the confluence of three forces: the Market, end-user Companies/Customers and vendors & service providers that offer print-related solutions.

Customer Influences

Traditionally, companies have been faced with *fragmented* Imaging & Output environments, meaning that different departments and business units have profit & loss responsibility for various aspects of the Document-related Information Supply Chain. And compounding the problem is the fact that user departments commonly have the autonomy to purchase "office supplies" (including printers and toner) up to a certain dollar amount — usually $1,000 — without requiring approval from I/T. The combination of fragmentation and end-user autonomy has contributed to a situation where companies have too much stuff that is unmanaged and untracked. The result: waste, high costs and inefficiency. One of the reasons why fragmentation exists (and has thus resulted in increased interest in MPS) is because of the introduction of MFPs into companies.

Also traditionally, the Information Technology (I/T) department has had responsibility for network-attached printers since the I/T department controls and managed the network. Because I/T "own" printing, they also assumed responsibility for their support and even supplies/consumables replenishment. When MFPs were introduced, the I/T department believed that the MFPs were nothing more than network-attached printers with the ability to copy and fax. So, I/T believed that they should own and manage MFPs, too.

On the other side of the house you have the Facilities/Purchasing/Real-Estate departments that have traditionally held responsibility for single-function analog copiers. With the introduction of MPFs they believed that MFPs were nothing more than analog copiers that now print and attach to the network, so they believed that they should continue to own responsibility for the procurement of MFPs and the associated supplies & consumables. And as for the network support, they agreed to abdicate this lone responsibility to the I/T department. This reality led to waste and duplication of effort (because companies now had two separate departments buying the same device twice for the same departmental needs), conflict between I/T and Facilities/Purchasing/Real-Estate (over the network and who should own the MFPs) and higher-than-necessary costs since the two departments despise each other so much over the issue that they never jointly plan for the acquisition and management of MFPs.

MPS: MANAGED PRINT SERVICES

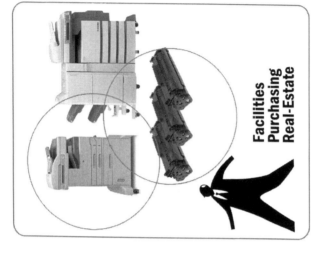

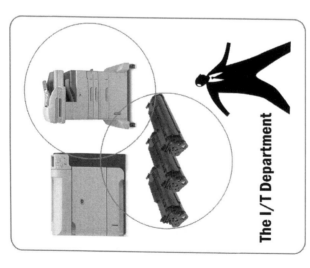

Over time, company executives decided that the in-fighting and waste were non-productive and that since the technologies of copiers and printers were converging into MFPs, so too should the company converge responsibility for copiers, printers and MFPs under one entity (see the diagram *The Shift: Convergence* below). Thus was born the converged Imaging & Output entity with responsibility for all Imaging & Output, including the DISC. And along with the convergence came a new role responsible for managing all Imaging & Output. Many people are familiar with the nickname of "Output Czar" for this role. Other job titles I have heard include Output Compliance Officer, Chief Output Officer and the title we introduced, the MIO — Manager of Imaging & Output.

As companies converged all I&O technology and management responsibilities under one entity, they have also begun to demand that their I&O vendors and suppliers work with them in a manner that is consistent with their convergence. In other words, companies have begun demanding that if a vendor wants to continue to do business with them, the vendor would have to provide *all* of the required Imaging & Output products and services under a converged (read: bundled) arrangement such as Managed Print Services. This requirement has forced many vendors and solution providers — often unwillingly — to enter the MPS business if they want to continue doing business with that customer or prospect.

The benefits promised by internal company convergence have not only led to increased interest in Managed Print Services by the end-user companies, but it has also sparked growth in the number of vendors that offer MPS solutions.

MPS: MANAGED PRINT SERVICES

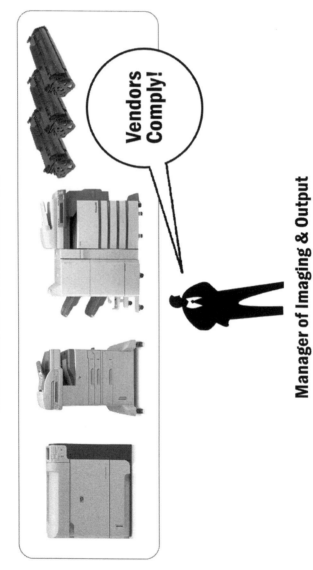

Market Influences

In recent years and especially since the recession took hold of the economy, companies have become increasingly focused on reducing costs, streamlining business processes, providing new services and optimizing I/T and Purchasing. These objectives are critical to a company's success. In a 2009 survey conducted by Infosys, clients were asked of their strategies and optimization plans in the short and medium terms. The results revealed that *cost reduction* is the most important factor driving sourcing decisions. About 50% of the respondents indicated that their near-term key driver for outsourced solutions (such as Managed Print Services) is cost reduction (see the table below).

KEY DRIVERS FOR OUTSOURCING	RESPONSES
Reducing Cost	**50%**
Gaining Competitive Advantage	18%
Lack of In-House Resources	15%
Leveraging Technology Expertise	0%
Availability of Cheaper Manpower	8%
Flexibility	8%
Consolidating Assets/Resources	8%
Source: Infosys 2009	

According to the poll, I/T organizations are encouraging value-based pricing models that enable both the client and the vendor to share the risk and reward, allowing them to work together toward the common goal of improving efficiency and bringing in predictability and flexibility. 55% of the respondents indicated that fixed-price managed services like MPS is the preferred pricing model to work with. Managed Print Services is an ideal option for enabling this type of risk-reward working relationship, while providing a proven vehicle for helping companies reduce hard-dollar costs associated with Imaging & Output.

Internal Vendor Influences

Hardware prices — and hence, the margins — on the sale of a printer, copier or MFP are under extreme pressure. The prices for printers continue to drop year-over-year, even to the point that resellers in the color laser printer space often make as little as $50 when they sell a printer. On top of that, resellers/VARs that sell on the aftermarket are being double-impacted by the decreased average "life" of a product installed with the first user. The trend is for end-user companies to depreciate a printer-type asset over 3-to-4-years as opposed to the traditional 5-year period. So going from 5-years of useful life to 3.5-years (on average) results in a 30% decrease in the expected aftermarket revenue opportunity for printers placed by resellers/VARS.

It goes without saying that hardware resellers/VARs cannot remain viable over the long-term by simply relying on the revenue generated from the sale of Imaging & Output hardware. In fact, some estimate that 35-40 percent of resellers/VARs will go out of business if they don't make changes. By focusing on the printer hardware, resellers may be missing an opportunity to capture an additional revenue stream: supplies & consumables.

Resellers and VARs that want to remain "viable" as solution providers will have to shift their paradigm from relying on the low-margin box-sale, to providing higher-margin bundled solutions and services. The one solution that is proven to deliver higher margins and provide to customers the higher-margin consumables (a $44B opportunity world-wide) and services is some form of a Managed Imaging & Output Service, such as Managed Print Services.

Managed Print Services offer the opportunity for a new source of revenue and a higher-margin revenue stream for 3, 4 or 5-years. As I discussed earlier, Imaging & Output margins have been shrinking for several years. For this reason, many dealers and resellers have turned to providing break-fix service and consumables in order to make up for the revenue short-fall plus growth. When you think about it, there are only a few ways to generate meaningful revenue from selling printers. One way is by pursuing a low-cost/high-volume strategy. Another way is by using the printer as a loss-leader and basically giving it away as part of, say, a desktop bundle that includes a PC, printer and software license. Another way to make money from selling printers is by selling the consumables, too.

According to Oxford Hill Consulting, the company/reseller/VAR that sells a printer to a consumer or a corporation only gets approximately 26% of the consumables spend of these companies for that device. In other words, *74% of consumers don't buy ink and toner from the reseller/VAR that sold them the printer.* Basically, resellers are doing the tough job of selling the printer and, the easy part — the consumables sale — is going to the stationers or to their competitors.

This is a significant lost opportunity when you consider that over the 48-month life of a printer, the total cost of the consumables can be anywhere from 10-to-30 times of that of the hardware. For example, if a color laser printer costs $1,600 and the cost for the CMYK toner cartridges total $1,040 at a 10,000 page yield, and there are 10 users of the shared-printer and the customer replenishes toner six (6) times per year, then over the 4-year life of the printer, the company will have spent $24,960 on consumables compared to $1,600 for the printer — a 15X differential.

Providing professional, value-added services to customers is another way resellers/VARS can earn incremental revenues at higher margins that offset the shrinking margins delivered by hardware sales. The advantage to VARs is that they can offer additional services (that translate into customer value) on top of hardware and consumables without any significant increased costs. Plus, it forces the seller to help the customer solve *business problems* through professional services rather than trying to shove boxes down their hallways.

Although margins on break-fix services have begun to decline (just as the margins on printer hardware), the margins on other types of printer-related services remain healthy because — in many cases — these services can be provided by existing personnel. Whether it's implementation services, fleet monitoring & management, transition planning, Imaging & Output assessment services, or supplies management, resellers/VARs are discovering new services revenue sources from existing assets.

THE SHIFT TOWARD MPS 37

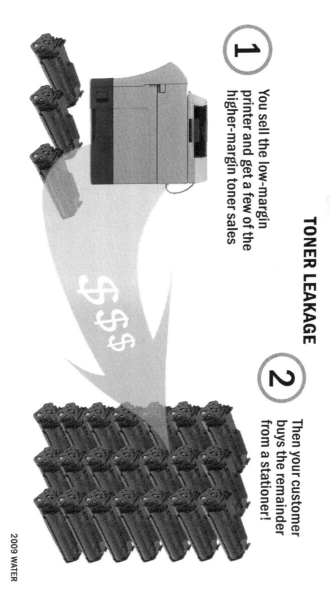

TONER LEAKAGE

1. You sell the low-margin printer and get a few of the higher-margin toner sales

2. Then your customer buys the remainder from a stationer!

2009 WATER

Competitive Advantage

Having a Managed Print Services solution available in your solution portfolio will give you an easily-understood advantage over a competitor that does not offer a Print Management, Managed Print, or equivalent Imaging & Output solution. Customers don't buy "printers," they buy the things that the printer delivers to them that solve a problem. In other words, customers want *solutions* to problems, and a printer in-and-of itself is not a solution. A solution (in its most basic terms) normally involves hardware, software and services, not just the *hardware*. So if you are in a competitive selling situation selling a managed solution against a competitor that is just selling a bunch of *printers* — from a TCO standpoint — the MPS-provider will have a stronger, more compelling offering to help customers accomplish their business objectives (e.g. reducing operational costs).

Customers buy things for peace-of-mind and security, and that is what MPS offers: a comprehensive managed solution with guaranteed service levels that enables managers to focus on their core competencies and not have to worry about the performance of its Imaging & Output operation.

The combination of these three forces — Customer Influences, Market Influences and internal Vendor Influences — have contributed most significantly, in my opinion, to the dramatic rise in interest and adoption of Managed Print Services on the part of both the customer and the vendor.

FOUR

The Elements of a Managed Print Services Solution

THE MANAGED PRINT SERVICES SPECTRUM: BASIC TO ROBUST

A Managed Print Services solution offering can vary in the degree of robustness and complexity, ranging from a very basic MPS solution bundle of printers, ink & toner, break-fix support and device monitoring, to a robust MPS solution that includes all of these elements listed in the next section (Elements of a Robust Managed Print Services Solution).

Just as there is confusion about the definition of Managed Print Services, so too is there confusion about what *really* constitutes a MPS solution and what can truly be considered a managed print service. As I wrote earlier, I believe that for an offering to be considered a "Managed Print Services" solution it should come in the form of a bundled solution that requires the service provider to take primary responsibility for meeting the customer's office printing needs through a combination of hardware, supplies & consumables, service & support and overall fleet management. Based on this description, for any print services offering to be consider a MPS solution it must — at a minimum — consist of a solution bundle of the following:

1. *Printers (including MFPs).* It can't be called a managed *print* solution unless it involves a device for creating printed output. This begs the question: does a MPS solution have to include the provisioning of new printers? My answer is, "no." However, if the customer has purchased printers that they do not want to replace, then in order for the service that a vendor provides to be considered a Managed Print Service solution, the vendor must include the *management* of that existing fleet into its service bundle.

2. *Ink and/or Toner.* The service provider must include a bundled offering of everything required to enable the customer to produce printed output, and that includes the supplies & consumables — and at least the ink and toner.

3. *Break-fix Support.* In order for the customer to produce printed output the devices must be operational, so the solution provider must ensure that the customer's printers/MFPs are covered under a service agreement that provides for the repair or replacement of the devices when they underperform or fail.

4. *Fleet Management.* Fleet management (at a minimum, tracking the status of the printers and/or the page volumes produced on them) is necessary to help relieve the customer's burden of managing their Imaging & Output environment. It is also helpful in tracking the number of pages produced on each device for billing and administration purposes.

At a minimum, in order for a solution to be considered MPS, it must meet these four criteria. Otherwise, the vendor is providing something less, such as a simple print services solution or even just the standard fare transactional solution.

Elements of a Robust Managed Print Services Solution

Site Assessment
In order to baseline a company's current state of operations and determine the level of waste, excess capacity, inefficiency, high cost and — most importantly — how well its current performance matches expectations — it is necessary to so some level of an assessment. An assessment is a systematic approach for gathering data to determining how well a company's Imaging & Output environment is operating compared to how it would like its Imaging & Output environment to operate. The assessment reveals the company's opportunity for improvement and cost reduction.

Solution/Fleet Design
After gaining an understanding of how the company's Current State environment looks and operates, it is then possible to develop an informed solution (including the recommended device fleet and a device deployment morel for efficient user-related workflow) to the problems that the Assessment revealed. The solution/fleet design should result in user efficiencies and, hopefully, operational cost reductions.

Hardware
Printers, MFPs/copiers, Fax machines and print server appliances and equivalent servers.

Supplies & Consumables

Ink, toner, ribbons, paper, preventative maintenance/maintenance kits, drums, fusers, rollers, wipers and others.

Software & applications

Fleet monitoring/management software, scan-to software, usage tracking software, LAN-fax software, print server software and others.

Break-fix Support

Remote diagnostics, on-site support, on-site Administrators, multi-vendor support, regional and national coverage.

Preventative Maintenance

Includes monitoring the device fleet to identify problems before they happen and/or upon occurring; developing contingency and resolution plans in the event of a device incident; installing preventative maintenance kits to help maintain device service levels.

Discovery and Design

As part of the MPS roll-out process, the solution provider will provide resources to go on-site to the customer's location (on a floor-by-floor basis) that is target to be installed. These resources will conduct a mini-Assessment to understand what devices are currently installed and what a good replacement fleet/deployment design would be for that floor. When a new design is agreed upon, the devices will be installed on the designated floor.

Installation

Includes transport from the loading dock to the staging area, device staging, un-boxing, inventory & tracking, hardware set up, connection (A/C and network), Configuration (IP, N/W, Subnet, gateway), testing and driver provision & related services.

ELEMENTS OF A MPS SOLUTION

De-installation

Includes device removal and relocation, hard drive disposition (removal, erasure, etc.) and packaging & re-boxing for shipment.

Equipment Disposition and Recycling

Some solution providers and third party companies offer customers the option of having them properly, legally and environmentally responsibly dispose of their old equipment — for a fee.

Existing (Multi-vendor) Fleet Administration

Includes managing a company's existing fleet of devices, regardless of the make, model and manufacturer of the devices. Some solution providers will limit this service to devices of a certain age (e.g. more than 2-years old) or even of a certain manufacturer. Others will buy them back from the company (so that the service provider owns them) and roll them into the company's Managed Print Services contract.

Deployment project management

Providing a Project Manager or a PM methodology or template to facilitate the project management aspect of a Managed Print Services engagement. The Project Manager is responsible for defining the project's deployment scope and building the project/deployment schedule. He/she is also responsible for ensuring that what was agreed to is delivered.

Fleet Management

Monitoring and managing the hardware fleet, including device usage, device tracking, remote administration, moves (including device re-locations), adds, changes and ensuring hardware service levels are met.

Asset Management

Can involve using software tools to help you keep track of not just what hardware and software you have running, but when you purchased each machine, what firmware each is running, who installed what on the devices, how heavily they're being used, expiring and redundant leases and when & how the devices will be (or have been) retired.

Supplies Replenishment
A varied service offering MPS customers the ability to have supplies & consumables replenished automatically, implement a supplies ordering process, various delivery options (dock, desk-side, etc.), toner installation and on-going supplies management.

Education & Training
Training ensures that the users of the (new) fleet of MPS devices know how to use them for effective exploitation of the devices. If users don't use the full functionality of the devices (such as MFPs) they you have basically wasted money on a set of features & functions that go unused. Training can also include helpdesk training, fleet tracking tool training, software training and more. Training delivery options for MPS solutions can include on-site training, Train-the-Trainer (T3) training, Web-based training, job-aids and other user documentation.

Transition Management
Transition Management involves developing a plan to get from the Current State to the Future State. But when we think of Transition management, we typically think about making sure the users make the transition from the old fleet deployment and device types to the new deployment model, including the ability (and comfort level) to effectively use and exploit the new technologies. Transition management and

Training go hand-in-hand.

One of the major reasons why MPS solutions fail to deliver on their promises of improved user efficiency, improved user satisfaction and even cost savings is because the users never adopt the new MPS infrastructure of devices and, therefore, never use them in the manner in which they are intended to be used. Therefore, an effective Transition & Communication plan is critical.

Usage tracking (for Billing and Optimization)
For most Managed Print Services implementations, the solution provider will provide a mechanism for tracking the number of pages that are produced on the hardware fleet being managed under the MPS agreement. This usage tracking is often required for billing purposes and for determining whether the devices are being used efficiently in their current physical location.

Ongoing Fleet Optimization
Ongoing fleet optimization involves tracking the usage (page output volumes and types) of each device against the number of users who are using the device, and then making adjustments (upgrades or downsizing), or relocation decisions to ensure that each device in the environment (those under the MPS agreement) are being used efficiently — meaning, a sufficient volume of pages are being produced on the device each month resulting in an average cost-per-page that is "reasonable."

Flexible Payment Options and Single-Invoice Billing
One of the attractions of the way many Managed Print Services solutions is offered is the option to acquire the solution with no initial capital investment. Under this approach, customers will not be billed for the solution until devices have

been installed and users have begin to use them; this is attractive for companies with cash-flow issues. It is also attractive because — when performing a cost-benefit analysis of the MPS proposition — it delivers a more attractive Return on Investment (ROI) and a shorter break-even period because there is no initial cash investment required.

Managed Print Services solutions traditionally offer a variety of billing options, including single-monthly invoicing, electronic invoicing, invoicing by department (and in some cases integration with in-house billing applications). And the flexible payment options include:

1. *Base + Click.* A fixed monthly base charge per device plus a variable charge based on the number of pages printed. This is the most common MPS billing method and should provide the lowest overall price.

2. *Pure Per-Page Pricing.* Companies are charged a cost-per-page for each page printed and no fixed monthly fee per device. With this pricing method customers are only charged for the pages they print which makes this financially risky for the solution provider because there is a chance that the customer will not print a sufficient quantity of pages to cover the solution provider's costs. As a result, this pricing approach results in a higher price to the customer because the solution provider must factor in the inherent risk of this pricing model.

3. *Level Estimated Payments.* This approach is also referred to as a Level-Pay approach wherein the solution provider charges the customer a fixed monthly fee every month for 12-months based on the anticipated number of pages the customer is anticipated to print (based on the results of an assessment). At the end of the 12-month period the solution provider will calculate the actual number of pages the

ELEMENTS OF A MPS SOLUTION

customer printed and adjust the customer's fixed monthly fee up or down (based on whether the customer printed more or less pages than expected) for the next 12-month period.

4. *Pre-paid Pages + Overages.* Under this approach, the customer will purchase a certain number of pages up-front (based on anticipated usage) and for any pages that the customer prints above that estimated number of pre-paid pages, the customer will incur an overage charge which is typically higher than their pre=paid page price. This is probably the least used payment option.

Technical Phone Support

Phone support can include Level -0 support (a call placed to an on-site MPS administrator before the call goes to the company's normal I/T helpdesk) all of the way through Level's 1 and 2 helpdesk support.

On-Line Customer Service Portal

Some solution providers provide a web-based service portal for their MPS customers which gives the customer the ability to make service requests, look up account and administrative information, order consumables and check their billing.

On-going Service Management

The MPS solution provider will assign a designated contact person (MPS Service/Contract/Engagement Manager) for their MPS customer as the primary interface for any MPS-related issues. Often, this MPS resource will conduct monthly and quarterly reviews with the customer to review the MPS program and how it is working.

Change Management

Services that include installs, moves, adds and changes.

Contracts and Agreements

Many MPS solution providers offer flexibility in their agreements (Master Service Agreements, Customer Service Agreements, Statements of Work, etc.) to include the ability for customers to adjust the MPS device fleet up or down by some percentage each year with no penalty.

Security

Things to consider are print output, network sniffing, device theft, hacking & re-routing, and network penetration.

FIVE

The Managed Print Services Process

THE MPS SOLUTION PROCESS FLOW

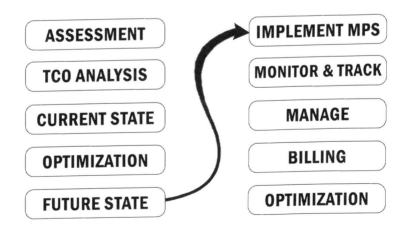

The Assessment

In order to understand the magnitude of a company's Imaging & Output-related problems, the company must first determine whether or not they actually *have* a problem and, if so, just how bad the problem is and if it's worth doing anything about in the short-term.

The most effective way for a company to determine if they have a problem (of a magnitude that warrants corrective action) is to *understand* the area of consideration and determine how far the condition that exists in the environment deviates from (or provides an opportunity for) satisfying the company's/departments/division's business objectives. And the most effective way that I have found to gain a thorough understanding of a company's Imaging & Output environment is by performing a detailed analysis or *Assessment* of the environment.

An assessment is a systematic approach for gathering evidence on how well performance matches expectations and standards; it involves analysis and interpretation of the evidence and using the resulting information to document, explain and improve performance.

A Managed Print Services/Imaging & Output Assessment can be simple or robust, depending on the scope of the initiative and the type of information that is required for the company or the solution provider to conduct a meaningful analysis of the environment under consideration. For a simple Assessment, data gathering can be done by asking the customer for insight, researching on-line, using various data-gathering methods (e.g. data collection or print tracking software), or even by extracting information from a printer configuration page to gain intelligence on Imaging & Output usage.

A more robust Assessment typically involves gathering more detailed, varied, complex information about the environment and using the information to understand current performance (device usage, page volumes and printing costs), how it compares to the company's desired performance and the company's opportunities for cost savings and improvement.

In late 1998/early 1999 while I was a Senior Consultant at Hewlett-Packard, I developed a simple-yet-effective Assessment methodology that has since been adopted by major hardware manufacturers and consultancies alike. Instead of using print tracking software to approximate monthly page volumes, I arrived at an approach whose results are exponentially more statistically-significant (read: reliable) than the approach whereby the solution provider collects data for a period of 30 or 60-days using print tracking software/tools. I coupled this usage calculation method with a data/cost analysis tool and an approach that combine to deliver analysis data that is as reliable and fast as can be expected using a sampling methodology. (For a detailed description of this methodology and a comparison of the different data gathering/page volume estimation processes, please refer to the book *Paper Problems* by Tab Edwards, Oxford Hill Press).

The process I developed (formerly called the "Hardcopy Operational Assessment — HOA at the time) follows an approach outlined as described below:

Define the Purpose for Engaging in the Assessment Initiative

The Purpose defines why the initiative is worth pursuing and how you ultimately expect to use the information that results from an Assessment. Failure to define a Purpose for the engagement could result in the project team spending lots of time engaging in activities that don't support the business Objective that the team is working to accomplish. In other words, the team could waste a lot of time. In addition, by defining the project's Purpose you will facilitated the strategy development process by providing focus for the team's efforts.

Determine a Representative "Study Group"

It is impractical, costly, time-consuming and even wasteful to assess or study an entire company, depending of the size of the company. The reality is that by using a sampling methodology, a company can study a subset of the company (the "study group") and extrapolate the findings to represent the whole company with a reasonable degree of accuracy and a big degree of cost savings and time savings. For example, if yours is a 10,000 person company with 10 offices across the United States, there is no need statistically to conduct an assessment of each of the locations in order to get accurate results. Instead, a company can sample a *representative* subset of the organization (a subset that consists of the various usage characteristics that exist in the company, such as accounting, I/T, marketing, administration, etc.) and get results that are perfectly usable and representative of the actual performance.

Collect Quantitative & Qualitative Data

The most important stage of the Assessment process could be the Data Gathering stage. Both the quantitative data (numbers-based data including page volumes and costs) and qualitative data (operational, infrastructure, user-related, workflows, etc.) serve as the foundation upon which the entire assessment is built. The Data Gathering activity (or at least some portion of it) is normally performed on-site at the company's study group office location(s).

Collect User Feedback

An important part of any plan, strategy, project, or Assessment is the involvement of the end-user community. Part of every Assessment should collect feedback from the users to get their input on what works, what doesn't work, what's missing and what they believe they need in order to make them more productive. Getting user involvement up-front is important for ensuring user buy-in when the solution is implemented. It is provides the company with valuable information about how to improve the users' Imaging & Output experience.

Develop Topological Map

Identifying where each device (printer, copier, MFP, fax machine) and where the users physically exist on each floor in the study group helps with the solution design stage of the process by identifying where users sit in relation to the devices they use in the course of doing their jobs and the likely user-related workflow process (their process of printing to a device and returning to their desk/workplace).

Perform a Workflow Review

As mentioned above, this is workflow from a user's perspective, meaning what is the process they go through to use the Imaging & Output devices in the course of doing their jobs. How much time do the spend wasting time "walking paper?" Understanding these user-related processes will help you design a solution that considers user productivity improvement.

Collect Cost Data to Determine Current TCO

Total Cost of Ownership (TCO) is the total cost of acquiring an asset, owning it and making it available to users over an extended period of time. When determining the *true* cost of Imaging & Output as part of an Assessment, you must take into consideration all of the costs associated with Imaging & Output.

Analyze the Current State Environment

Once all of the data have been gathered, it is then possible to create a Current State snapshot or benchmark of how the company looks today, including devices, page volumes, costs, user-related workflow, user feedback and existing areas of waste, high cost and inefficiency — and their magnitude.

Design a Solution for Improvement (the *Future State*)

Once you have identified the areas of waste, high cost and inefficiency, it is then possible to design a solution that would reduce hard costs, eliminate waste and improve user efficiency and satisfaction.

Develop a solution design model for other locations

If you did not study the entire company, then you can extrapolate the findings from the study group to represent the entire company. This will give you a sense of how bad the problems are today company-wide and what the overall opportunity for improvement and cost savings are.

TCO Analysis

As stated above, the Total Cost of Ownership is the total cost of acquiring an asset, owning it and making it available to users over an extended period of time. It is important for companies to understand that you are likely paying more for Imaging & Output than you imagine. When I conduct planning sessions with company executives (who know very little about the particulars of Imaging & Output and I&O TCO), in order to get their buy-in on the different cost factors that I will use to determine their TCO, I ask them the following question as a means of getting them to understand that they are spending money on more things than they probably imagine: *If you didn't have any printers, copiers/MFPs, fax machines, or scanners in the company, which costs that you incur today because you have these devices would you no longer have to pay?* The list should look something like this:

- The purchase price or lease costs of the devices
- The cost of paper
- The cost of supplies and consumables
- Maintenance costs
- Power consumption costs
- Phone line, port and per-call charges
- Network port costs
- Printer-related network management costs
- Installation costs
- Print servers and software license costs
- Helpdesk costs
- Asset management costs
- Floor space/real-estate costs
- Personnel costs/key operator costs
- User training costs
- Various productivity and soft costs
- etc ...

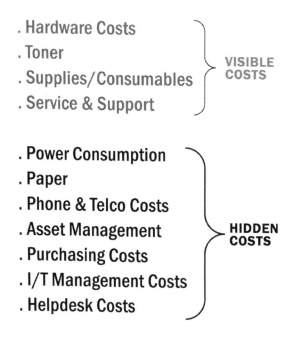

Optimization

The solution design you develop should be an optimized version of the company's existing environment that eliminates redundancy, excess capacity, inappropriateness-of-fit between user needs and devices, streamlined workflows and cost reduction.

Monitor & Track / Manage

(Ideally) using a fleet tracking or fleet management tool, monitor each device for preventative maintenance and fleet administration. Depending on the needs of the company, you may also wish to track device usage and page production down to the individual user level. Why a company would want to track usage down to the individual user level (e.g. how many color pages did Jane print this month?) will depend on a need.

Billing

One of the financial attractions of Managed Print Services is that vendors often provide these solutions to customers with no initial capital outlay required, and the MPS-buyer only pays for the solution when they start using the devices. Then, the solution provider will bill the company on a scheduled basis — usually monthly. This is a convenience that companies enjoy about MPS; it's easy to have in place and to pay for.

SIX

Total Cost of Ownership vs. Cost-Per-Page

What Companies Want

One can make a strong argument that virtually all for-profit companies essentially want five things: profit growth, new business, loyal customers, productive/happy employees and a sense that they are doing something socially-responsible. Many organizations believe there is an interrelationship between these elements that drive profitability. In one interesting case, in fact, members of the Sears executive team realized that there is a chain of cause-and-effect running from employee behavior to customer behavior to profits ("The Employee-Customer-Profit Chain at Sears" in *the Harvard Business Review*, Vol. 76, No. 1 January 1998). But any way you slice it, and as noble as social responsibility sounds, most for-profit companies are focused on profit growth.

In its simplest form, profit is total revenues less total costs/expenses. The question is: how can Managed Print Services help companies improve profitability? The answer: by helping companies reduce operating costs; this is what companies are interested in and, in many cases, this is what companies seek assistance to accomplish. The strong MPS seller will understand this and will look for opportunities to help their customers improve profitability. This is what "consultative" selling is all about – helping customers accomplish *their* business objectives. So the strong MPS seller will focus on helping the customer reduce operating costs in an effort to help the company improve profitability. And the *winning* MPS seller will focus on helping the customer reduce ALL Imaging & Output-related operating costs – Total Cost of Ownership – to help the company *maximize* profitability.

Cost-Per-Page (CPP)

Simply put, cost-per-page (CPP) is defined as the total cost of an asset divided by the number of pages printed/copied on that asset in a given month. So, for instance, if a printer costs $100 per month to own and operate and that printer prints 100 pages per month, then the average cost-per-page for this printer (given its current usage) would be $1 per page; as I explain in detail in my book *Paper Problems*, this is how the exact same device placed in 2 different settings can have wildly different costs-per-page.

Cost-per-page is also a relative unit of measure, meaning, considering one thing relative to another – a comparison. Cost-per-page is the lowest common denominator when it comes to comparing one type of Imaging & Output technology to another. For instance, if a company wants to know whether a single-function printer is more cost-efficient as compared to a MFP and an analog fax machine, the company can compare the average costs-per-page of the pages produced from each device to determine which device is relatively the most cost-effective. By definition, cost-per-page is a measure used for relative comparison, and when it comes to providing customers with cost-per-page price quotes, CPP is used to compare the prices of one proposal over another. Between 2008 and 2010, the consultants at Water have polled customers that have asked for CPP quotes from vendors to understand why they were interested in receiving a cost-per-page quote. In approximately 90% of the cases, the customers stated that they wanted the CPP price quotes to compare the relative prices of the competing proposals and to use the information as leverage to negotiate a better price.

This reality is one of the primary reasons why many experienced MPS consultants and sellers are reluctant to engage in cost-per-page discussions with prospects and customers. Another reason is because they realize that a discussion about cost-per-page does not help the customer accomplish a cost-reduction business objective aimed at improving profitability – the main objective that most for-profit companies want to achieve.

The Problem with Cost-Per-Page from the Buyer's Perspective

If yours are a company looking to implement (buy) a Managed Print Services solution and your focus during the proposal evaluation stage is on securing the lowest cost-per-page offering from various vendors, then it is possible you are missing an opportunity to reduce your company's operating costs to the greatest extent possible. In most cases when vendors provide customers and prospects with cost-per-page quotes, the vendor only includes the cost of the hardware, ink/toner and service in the price bundle. So, for instance, if the company prints 10,000 pages per month at a current cost-per-page of $0.03, then the company is spending $300 per month for this device bundle. Now, say, the customer gets a cost-per-page quote of $0.02 per page to replace the company's existing MFP device, the company will now only be spending $200 per month, for a total cost reduction of $100 per month or $1,200 per year. Not bad. And this cost reduction will definitely help the company reduce its overall operating costs and improve profitability. I can see it now: this customer runs to the Chief Financial Officer (CFO) with visions of a company-sponsored cost-cutting award in mind and proclaims that he/she has saved the company $1,200 per year, to which the CFO replies with a tinge of disappointment, "Is that *all*?" Now suppose this CFO was faced with the choice of investing in this MFP project that saves $1,200 or investing in another project that saves $2,000 – and the CFO's interest is in saving the company the most money. The CFO would likely choose the project that saves the company $2,000, meaning no new MFP and no cost-cutting award. The reality is that the customer's $1,200 cost savings is actually more like $1,800 in savings when you consider all of the associated costs that could be reduced by the introduction of a new MFP device.

Traditionally, when companies deal in cost-per-page price quotes they are creating an incomplete picture of the company's cost-reduction potential because the only costs considered are the "basic 3": hardware, ink/toner and service. Are these the only costs that actually get reduced when you upgrade, migrate or optimize a hardware fleet? Of course not. But far too often when companies request price quotes from sellers they only consider the "basic 3." Why? Because this is a remnant from the analog copier days when copier dealers sold their copiers to customers bundled with maintenance and toner – all for a tidy cost for each page they produced under the terms of the lease agreement. This was at a time when copier dealers/sellers didn't have MFPs to sell, they didn't consider printers and they were not as sophisticated in their understanding of all of the costs a company actually incurs when they purchase and install printers, copiers and MFPs. But now that we are far more knowledgeable about all of the costs that a company incurs and can reduce by upgrading, migrating or optimizing its hardware fleet, we realize that only considering the costs of the "basic 3" is creating an incomplete picture of a company's *true* Imaging & Output costs and potential savings.

When I consult with companies that are pursuing the purchase of MPS solutions, and whose focus is on securing cost-per-page quotes from multiple vendors, one of the first things I ask during the planning process is this: Why do you want a cost-per-page quote? From that point, the conversation typically goes something like this:

Customer: "Because we want to compare the different quotes apples-to-apples."

Me: "Why?"

Customer: "So that we can decide which price is lowest."

Me: "Why do you want a lower price?" (At which point they look at me as if I was insane)

Customer: "To save money."

Me: "Ohhhh ... so you want to *save money*. Then let's focus on saving you *money* and *not* on the vendors' cost-per-page numbers."

Think about it. Customers don't *really* want a lower cost-per-page. That's like going to the doctor and telling the doctor you want a prescription for *cough medicine*. You don't *really* want cough medicine, what you *really* want is to stop coughing. The same logic holds true when discussing cost-per-page. Astute customers acknowledge they don't *really* want a cost-per-page quote or even a lower price; what they really want is to *save money* and reduce costs to improve profitability. The focus should be on reducing costs and not on getting a lower cost-per-page quote. There is a difference. Just like there is a difference between wanting cough medicine and wanting to stop coughing. If the doctor focused on the price of different cough medicines and neglected to focus on why you are coughing and ways to stop your cough, she would be doing you a disservice.

The Problem with Cost-Per-Page from the Seller's Perspective

If you sell MPS solutions and you only provide your customer or prospect with proposals focused on cost-per-page, then you – just like the neglectful doctor referenced in the preceding paragraph – are doing your customer a disservice. How? By not helping your customer "stop coughing" ... uh, I mean ... improve profitability (or whatever their ultimate Objective). You are the MPS consultant and customers – who are not themselves MPS experts in most cases – rely on you to bring your MPS expertise and experience to bear on their problems. The expectation is that you, the consultant/seller, will educate the customer on ways to help them accomplish their objectives through the implementation of a MPS solution if it is possible to do so. One of your responsibilities as a MPS consultant is to help your customer understand their true costs of Imaging & Output and their true potential for operating cost reduction leading to improved profitability. Failure to do this could hurt one's credibility, and the most important trait of any sales representative or consultant is their credibility.

But selfishly, focusing on delivering a cost-per-page quote could be bad business. Customers mostly request cost-per-page quotes in order to compare prices and to use the quotes to get better pricing from the vendors. And that's okay; it's called negotiating. But for you as a MPS seller, this should not be your preferred path since, in the end, your customer will have negotiated your per-page price down so low that your margins will erode and you will end up "winning" a deal that you wish you hadn't.

MPS sellers and consultants must change the level of the discussion from cost-per-page to maximum cost reduction and profitability if they are to truly provide value to their customers. In order to understand maximum cost reduction, you must first understand the concept of Total Cost of Ownership.

Total Cost of Ownership (TCO)

As described in Chapter Five, Total Cost of Ownership (TCO) is the total cost of acquiring an asset, owning it and making it available to users over an extended period of time. When determining the *true* cost of Imaging & Output you must take into consideration *all* of the costs associated with Imaging & Output. It is important for companies to understand that you are likely paying more for Imaging & Output than you imagine. When I conduct planning sessions with company executives (who know and care very little about the particulars of Imaging & Output and I&O TCO), in order to get their buy-in on the different cost factors that I will use to determine their total costs and opportunity for operating cost reductions , I ask them the following question as a means of getting them to understand that they are spending money on more things than they probably imagine: *If you didn't have any printers, copiers/MFPs, fax machines, or scanners in the company, which costs that you incur today because you have these devices would you no longer have to pay?* They generate a list that looks something like this:

The purchase price or lease costs of the devices
The cost of paper
The cost of supplies and consumables
Maintenance costs
Power consumption costs
Phone line, port and per-call charges
Network port costs
Printer-related network management costs
Installation costs
De-installation costs
Disposal costs
Print servers and software license costs
Helpdesk costs
Asset management costs
Floor space/real-estate costs
Personnel costs/key operator costs
User training costs
Various productivity and soft costs
etc ...

In other words, these executives acknowledge that they do, indeed, spend money on more than just the "basic 3" of hardware, ink/toner and service. In turn, they also acknowledge that their opportunity for reducing operating costs far exceed the cost reductions generated by getting a lower cost for the "basic 3." For this reason, MPS sellers should be compelled to discuss Total Cost of Ownership with their customers, and customers should only be interested in having discussions about the totality of their Imaging & Output costs. Only then can you begin to conceptualize the potential for maximizing profit growth based on the operating cost reductions possible from your Imaging & Output environment.

What's interesting is that in all of the 13-plus years that I have been working with customers on Imaging & Output optimization initiatives, I can't say that I ever recall a time where a C-level executive from a major company ever wanted to talk about cost-per-page; they all want to talk about reducing costs and improving profitability.

CPP Relative to TCO: Considerations

Consideration	Cost-Per-Page (CPP)	Total Cost of Ownership (TCO)
Typical Implication	Selling a printer or copier/MFP	Selling a Business Solution
Customer Interest	Shopping for the Lowest Price	Maximizing Operating Cost Reduction
Ultimate Focus	Saving Money on the "Basic 3"	Accomplishing a Business Objective

SEVEN

Drawing Distinctions

When it comes to Imaging & Output in general and Managed Print Services in particular, there is often confusion about the distinction between the various print management solutions and the related elements of the Document-related Information Supply Chain (DISC). For instance, is there a difference between "workflow" and "process flow"? What about "Managed Print" and "Total Print Management"? To help clear up the confusion, I have provided some definitions for and distinctions between many of the most commonly-misused terms in Imaging & Output.

Transactional Sale

This is the individual transaction or simple "box sale" where the buyer purchases individual products (e.g. printers and toner) that are not part of a bundled Managed Print Services solution.

Print Services Solution

A print service is a combination of products and/or services that are sold together in support of a company's Imaging & Output environment. An example would be a company purchasing toner and service contracts together, or printers with toner together from a single solution provider. Though Print Service Solutions involve the sale of bundles items, they do not contain the minimal set of elements required for a solution to be considered Managed Print (hardware, consumables, break-fix support and fleet tracking at a minimum).

Managed Print Services (also referred to as Print Management)

As I wrote earlier in the book, a true Managed Print Services solution offering is typically focused on the General Office and requires the service provider to take primary responsibility (which typically involves fleet ownership) for meeting the customer's office printing needs through a combination of hardware, supplies & consumables, service & support and overall fleet management.

Workflow

"Workflow" is a term that is thrown around freely, often without consideration for the various interpretations of the word. The result is that *Person A* will use the word "workflow" and mean something totally different than *Person B* is thinking when they hear it. This disconnect can result in an inappropriate solution being provided for a given set of requirements.

So what is "workflow"? A workflow is a sequence of connected steps. It is a depiction of a sequence of operations, representing the work process a person, group of people, department, organization, etc. go through in the course of accomplishing a task. As you can imply from this broad definition, the term "workflow" can mean almost anything. That is why it is important to clearly define the manner in which you use the term in the business context, especially when it related to Imaging & Output and the DISC.

Document / Content Management

Document management, often referred to as Document Management Systems (DMS), is the use of a computer system and software to store, manage and track electronic documents and electronic images of paper based information captured through the use of a document scanner. The term document is defined as "recorded information or an object which can be treated as a unit". DM systems allow documents to be modified and managed but typically lack the records retention and disposition functionality for managing records. Key DM features are:

- Check In / Check Out and Locking
- Version Control
- Roll back
- Audit Trail
- Annotation and Stamps
- Summarization

Enterprise Content Management

Enterprise Content Management (ECM) is the strategies, methods and tools used to capture, manage, store, preserve and deliver content and documents related to organizational processes. ECM tools and strategies allow the management of an organization's unstructured information, wherever that information exists.

Total Print Management

Total Print Management (TPM) is a comprehensive approach to optimize and manage a company's document input and output environment fully, incorporating corporate-wide Imaging & Output devices, software, supplies, management and integration. TPM is all-encompassing across all Imaging & Output environments (see "Imaging & Output Environments" below) and can include document and content management; Managed Print Services is a subset of TPM.

Imaging & Output Environments

Companies generate lots of information, create lots of paper output and spend millions of dollars on printers and print services. Where does it all get produced? How much of the volume of paper created annually is created in the general office? The central reprographics department? The data center? How much is sent outside to print service providers? Where is the percentage of spend for these categories of output going? To answer these questions, we first need to define each of these segments based on the context in which I will use them throughout this document.

Home-Based and Satellite Office

Professionals such as sales representatives, consultants, Insurance agents and others who primarily work from a home office or a small satellite/smart office and thus require I&O resources at the remote locations.

Distributed/General Office

The distributed (general) office is the part of the organization where the end-users create, print, copy, fax, scan and finish (e.g. collate, staple) the pages and documents they use in the course of doing their jobs. End-users typically send output jobs from their PCs to the self-service printers and workgroup MFPs located throughout the general office environment.

Production/Central Reprographics

The production environment (also referred to as the *Central Reproduction Department*/CRD, the print shop and the copy center) is the in-house part of the company where high-volume document production and duplication is performed. Sometimes, the mailroom function is a part of this environment.

Companies encourage users to direct their print and reproduction jobs of certain sizes to the appropriate environment. For instance, some companies will instruct users to create output jobs of 100-pages or less in the general office, while sending larger, more complex jobs to the production environment. Normally, the high-speed (90ppm plus), high-volume MFPs used in this environment require an operator to run them.

Data Center and Mainframe

The output that is produced in the data center off of mainframe (large) computer systems is normally created on "green-bar" (impact) printers and high-speed cut-sheet printers. End-users rarely have access to these printers for use in creating their day-to-day output. Mainframe printers produce large reports and system data that are used by data center managers and other I/T professionals. Many times, companies will tie several printers together (clustering) to achieve extremely high speed printer output.

Commercial/External Print

Commercial printing (also referred to as external printing and Print Service Provider/PSP printing) includes marketing collateral, brochures, booklets, annual reports, forms, labels and other "professional" print requirements. This is the category of printing that is outsourced and performed by printing companies (PSPs) that are not owned by the corporation requesting the print jobs.

An emerging sub-category of the external print spend category is Internet-based print service providers that allow end-users to electronically submit print jobs over the Internet to be fulfilled by 3rd party printers.

WHERE ARE THE PAGES PRODUCED?

The reality is that in most large corporations, there will inevitably be a percentage of printed output for which the company cannot account. No one seems to know where the pages went or came from, or how much money was spent on the "lost" pages (in the general office approximately 30% of all pages produced are either lost or discarded - MBI). But they all will agree that they produce and pay for some percentage of pages that they absolutely cannot account for.

Based on actual study data gathered from output assessments that I have performed over the years, in addition to 2006 research data provided by Oxford Hill Consulting (across multiple industries), a general break-out of pages (there is an assumption of a direct correlation between the number of pages produced in an environment and their associated costs) can be described as presented below. Please keep in mind that the percentage break-outs provided below are a summary across multiple industries and could differ somewhat from the percentages in your specific industry.

- Distributed/General Office = 20.2%
- Production/CRD = 5.1%
- Data center/Mainframe = 2.4%
- Commercial/External Print = 70.6%
- Other and/or unaccounted for = 1.7%

(Source: Tab Edwards studies and Oxford Hill Consulting, 2006)

THE 7-Cs OF IMAGING & OUTPUT

EIGHT

The 7 Cs of Imaging & Output

My research shows that every company (with at least 1,000 employees) that purchases and makes hardcopy technologies available to its users over an extended period of time are, at some point, affected and influenced by seven categories of issues:

1. Coordination
2. Copiosity
3. Cash-Flow
4. Course
5. Cognizance
6. Consumer
7. Community

I refer to these issues as *The 7-Cs of Imaging & Output*. Any company that is serious about understanding the condition of their I&O environment must consider their company in the context of each of these categories of issues before sustainable improvements can be made. The data from 12 years of assessments and studies show that when companies take a comprehensive look at their Imaging & Output environments and address the issues they face in each category, they have been more successful at reducing costs, improving overall efficiency, instilling order and improving user-related workflows, leading to increased levels of user satisfaction.

THE 7 Cs OF IMAGING & OUTPUT

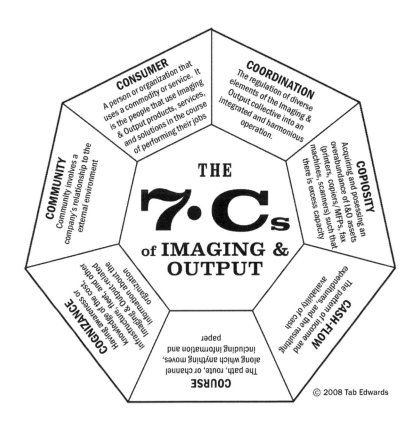

Coordination

Coordination refers to the regulation of diverse elements of the Imaging & Output collective into an integrated and harmonious operation. The concept of Coordination involves driving towards a common focus of resources, energies and attention in pursuit of the attainment of some set of objectives.

In most corporations the hardcopy infrastructure (the printers, copiers, MFPs, fax machines and scanners) has, until recent years, been an area of the organization that has not received a lot of focus. This is truer for the printer infrastructure than for the copier infrastructure. There are several reasons that Chief Information Officers offer to explain this condition, the most common being that the printers are inexpensive and they work, so in the grand scheme of things it's just not that critical an area to focus on.

This had been the prevailing attitude until corporate cost-cutting initiatives became regnant, forcing Information Technology (I/T) managers to look outside of the data center for other areas to generate cost savings. As managers began conducting assessments of their hardcopy infrastructures, they consistently found the following conditions to be true:

- Waste and high costs were worse than they had imagined
- The same, similar and functionally-overlapping devices and supplies were being purchased separately by the I/T, Sourcing & Operations (including Purchasing) and end-user departments, leading to a high concentration of devices per user, leading to excess capacity
- Devices were purchased without regard for a deployment strategy

- Corporate-wide device and cost information were mostly unknown to them
- There was an opportunity to significantly reduce costs and save money!

As companies began developing strategies to improve their hardcopy environments, one thing became clear: it will be very difficult implement an efficient, cost-saving solution and sustain the savings without a synergistic change in the organizational policies & procedures around requesting, purchasing, paying for, maintaining responsibility for and owning hardcopy assets.

Understanding the Problem

Since the mid-'90s — when spending on technology hardware was unprecedented — companies have become overwhelmed by the number of printers, copiers, MFPs and fax machines that proliferate throughout their offices leading to high costs and excess capacity. Today, we find that the primary causes for general high costs and inefficiency are fundamentally a result of the fragmented responsibility (dis-coordination), failure to manage and not taking advantage of improvements in technology. This chapter will focus on dis-coordination.

In most corporations multiple departments have authority, management and profit & loss (P&L) responsibilities that overlap in the general office hardcopy environment. For instance, the I/T, Sourcing and user departments may each have the ability to purchase printers, MFPs and supplies, while doing so under no overall Imaging & Output strategy. As a result, companies will have a high percentage of personal-use printers, "too many" devices of a similar type and multiple models of the same technology. The result is not only a duplication of effort, but duplication of costs, waste and excess capacity driven by: (a) dis-coordinated purchase, management and profit & loss responsibility; and (b) the lack of a single department or manager responsible for Imaging & Output company-wide.

When there is dis-coordinated or fragmented ownership and profit & loss responsibility, environments are generally inefficient and Imaging & Output costs are uncontrolled, leading to waste and an overall increase in Imaging & Output costs company-wide.

DIS-COORDINATED RESPONSIBILITY
DRIVING ENTERPRISE COSTS IN IMAGING & OUTPUT

DEPARTMENT	HARDWARE RESPONSIBILITY	HARDWARE PROFIT & LOSS	SUPPLIES RESPONSIBILITY	SUPPLIES PROFIT & LOSS	HELPDESK RESPONSIBILITY	HELPDESK PROFIT & LOSS	RECORDS MGMT. RESPONSIBILITY	RECORDS MGMT. PROFIT & LOSS	CRD/MAILROOM RESPONSIBILITY	CRD/MAILROOM PROFIT & LOSS	OUTSOURCED COPY RESPONSIBILITY	OUTSOURCED COPY PROFIT & LOSS	FAX RESPONSIBILITY	FAX PROFIT & LOSS	SUPPLIES STORAGE RESPONSIBILITY	SUPPLIES STORAGE PROFIT & LOSS	LOST DOCUMENTS RESPONSIBILITY	LOST DOCUMENTS PROFIT & LOSS
FACILITIES & OPERATIONS	○	○	○	○					●	●	○	○	○	○	●	●		
INFORMATION TECHNOLOGY	○	○	○	○	○	●												
DEPARTMENT MANAGEMENT	○	○	○	○	○		●	●			○	○	○	○			●	●
TELECOMMUNICATIONS					○		○	○					○	○			○	○
FINANCE																	○	○

● Areas of duplicate cost, duplicate P&L responsibility, and/or duplicate effort = lack of clear ownership

○ Areas of clear delineation

CHALLENGES

- Independent decisions > imbalance > higher costs
- Multi-source budgets > duplication > higher costs
- Multi-source budgets > overall costs ignored
- No single department can ensure optimal deployment
- Departmental "control" > uneven buy-in to solutions
- Inability to effectively roll-out corporate-wide initiatives

Acquisition, Support and Ownership Process

A discussion about organizational dis-coordination cannot begin until the common process most companies follow for acquiring, paying for, supporting and ultimately owning hardcopy technology is reviewed. The process a company typically follows can reveal a great deal about the degree of dis-coordination that exists and the areas of inefficiency (concern) inherent in the process.

Over many years of working with companies on strategy development and I&O initiatives, my research has shown that corporations typically follow the same or similar processes for acquiring, supporting and "owning" hardcopy assets. This process flow is described and illustrated below.

Scenario A: The typical process for hardcopy device acquisition by a user or department (This process is illustrated by "Scenario A" in the diagram below)

1. A business unit or a user has a "requirement' for additional hardcopy resources, features and/or functionality. For this illustration, we will assume the user has a requirement for printing resources.

2. If the user department has purchase autonomy — giving them the right to purchase equipment on their own within a pre-determined dollar amount; say $1,000 — the user will purchase the equipment (from the local electronics department store or otherwise) that conforms to the corporate standard. At this point, the user department often considers the device to be "theirs." [**This is an** *Area of Concern*]

3. If the user department does not have purchase autonomy, or the hardcopy device they require exceeds their

approved budget limit, the user department will follow the company's approved equipment request process to request the hardcopy equipment they "need."

4. If the request is for printers or MFPs, then the request is handled by the I/T department; if the request is for copiers (believe it or not, many companies still lease analog copiers) or MFPs, then the request is handled by the Sourcing department. If the request if obviously for a MFP, then the I/T and Sourcing organizations have to decide which is the appropriate group to handle the request. [**Area of Concern**]

5. The I/T and/or Sourcing organizations will take the request and try to "understand" more about the users' requirements in order to determine the appropriate device(s) to procure. Some managers will conduct an assessment of the requesting-department to gain a better understanding of the requirements; some will not. [**Area of Concern**]

6. If an assessment is conducted the manager will have a good understanding of the users' requirements and ill make a recommendation to the department. If an assessment has not been conducted, the manager will still make a recommendation to the user department which is not fact-based. [**Area of Concern**]

7. If the user department accepts the recommendation — and has the budget approved to pay for it — the process moves to the billing and accounting processes. If the user department does not have the budget, then the process stalls until funding is made available.

8. Once the device is installed, if it is a single-function printer or MFP, it will be managed supported by the I/T

organization. If the device is a MFP procured by the Sourcing organization, it will be managed by the Sourcing organization, but supported by the I/T organization (because of its network connectivity) and the hardware/services provider. [**Area of Concern**]

Scenario B: The typical process for hardcopy device acquisition as part of an I/T or Sourcing department initiative (This process is illustrated by "Scenario B" in the diagram below)

1. Whether initiated by a request from a user department or otherwise, the I/T and/or Sourcing organization will initiate a company-wide project involving a refresh of the hardcopy device fleet or a managed (or outsourced) I&O implementation. Oftentimes, such projects are initiated without the benefit of an up-front assessment or requirements analysis. [**Area of Concern**]

2. If the project is a corporate-wide initiative, the department(s) will assemble a task force to plan, manage and execute the project. The task force will develop a requirements definition to support the objectives of the project. Again, the requirements definition is developed without the benefit of a comprehensive, or even substantial, assessment. [**Area of Concern**]

3. In most cases, the company will issue a Request for Proposal (RFP) to multiple vendors requesting their solution to satisfy the requirements definition.

4. The task force will follow their pre-determined selection process to choose a solution provider. Once a solution provider has been selected, the task force will determine the desired acquisition method for the solution they have agreed to implement. Oftentimes, the acquisition meth-

od selected is based on past experience as opposed to the current realities of the market. [**Area of Concern**]

5. The task force will place an order for the equipment or solution.

6. If it is determined that the project will not be a corporate-wide initiative, the I/T and/or Sourcing organization(s) determine if there is a need for additional services to be included with the hardware acquisition. The responsible organization will then determine a department to pilot the hardware/solution. This determination is often made without the benefit of a requirements analysis in the designated pilot department. [**Area of Concern**]

7. Once it has been determined that a pilot of the hardware/solution is desirable, a well-developed business case is an available option to the task force. In most cases the task force will forego the development of a robust business case, even when the project has to gain approval and finding from the company's executive management. [**Area of Concern**]

8. If the project is not approved, it stalls. If the project is approved, the equipment/solution is ordered and delivered.

9. If the single-function printers and/or MFPs require network installation, the I/T organization is engaged and the installation project gets added to the I/T task list. Often, this request is made by the Sourcing organization for MFPs they acquired without the coordination of the I/T department. [**Area of Concern**]

10. The devices get installed and/or the solution gets implemented.

11. Training and transition management are performed with the user community.

12. Either the I/T department, the product/solution vendor and/or a 3rd party service provider will provide and be responsible for on-going support. [**Area of Concern**]

13. "Ownership" will reside with the I/T organization for printers and MFPs, or the Sourcing organization for MFPs. [**Area of Concern**]

THE 7 Cs OF IMAGING & OUTPUT

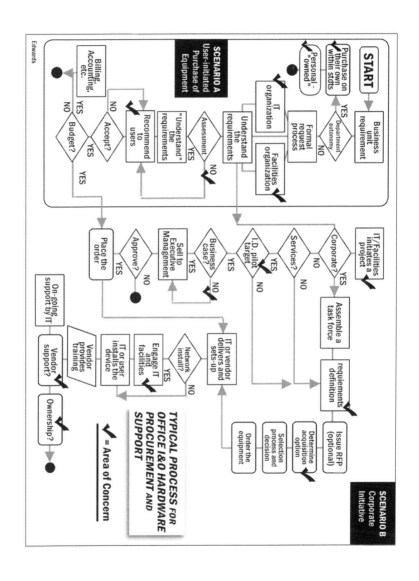

TYPICAL PROCESS FOR OFFICE I&O HARDWARE PROCUREMENT AND SUPPORT

◀ = Area of Concern

Internal Conflict Between the I/T and Sourcing Organizations

As corporations began introducing office-class multifunction devices/peripherals (MFPs), the potential for conflict between the I/T and Sourcing departments (along with the duplication of effort related to the acquisition and "ownership" of the devices) increased along with the devices' permeation throughout corporations.

In 1998 I was engaged by a major financial services company to facilitate a workshop designed to bring alignment and cooperation between the I/T organization and the Sourcing organization. The I/T organization held responsibility for the printers and the then-emerging MFPs, while the Sourcing organization also maintained control of MFPs in addition to copiers. Since MFPs attach the network and users can print to them, the I/T organization justified maintaining responsibility for them. Likewise, because the MFPs were copiers, too, the Sourcing organization also felt justified in maintaining responsibility for MFPs. Because of the un-planned dual-ownership situation, MFPs were being evaluated, purchased and brought into the company by both organizations, leading to multiple "standards," multiple vendors' products, multiple models, duplicate costs and duplicate efforts. But of all the internal challenges brought on by the emergence of MFPs into the company, the most serious was the in-fighting and the strain on the working relationship between managers in the two organizations brought on by the introduction of MFPs into the company.

Fast-forward 12 years. Even today, the majority of the *Fortune 500 Corporations* I meet with still struggle with the lack of alignment between I/T and Sourcing as it relates to the responsibility and "ownership" of MFPs and the conflict, lack of cooperation and coordination between the two organizations. This internal conflict is a drain on productivity and results in duplication of effort (oftentimes two separate Requests For Proposal will be issued, one from each organization, both requesting the same thing: MFPs) and devices. Though some I/T and Sourcing managers have taken it upon themselves to forge an agreement to make joint decisions about MFP acquisition and ownership, it still surprises me that most companies have not instituted formal procedures (such as creating a Chief Output Officer role or an Imaging & Output department responsible for enforcing standards and guidelines for all output acquisitions) for managing MFPs.

Excess Capacity

As implied above, a major problem of dis-coordination related to the acquisition if MFPs is that, although both the I/T and Sourcing departments can purchase MFPs (since they have the autonomy and P&L responsibility), the two departments often make purchase decisions independent of each other and the user departments. The result is excess capacity due to the duplication of MFPs (and the fact that more often than not, the MFPs are acquired in the absence of a deployment strategy or departmental assessment).

If the average annual Total Cost of Ownership (TCO) of a department-class MFP is $5,500 and the MFP-to-user ratio (in a 10,000 person corporation), is 1-to-3.4, then the cost of waste and excess capacity in MFPs could be more than $2.8 Million annually.

Contributing to the inefficiency is the fact that both the user departments and the I/T department purchase printers without any significant joint planning or requirements definition. This leads to additional waste and high costs. For instance, if the average annual Total Cost of Ownership (TCO) of a supported networked (shared) printer is $3,500 and the shared- printer-to-user ratio (in a 10,000 person corporation), is 1-to-10, then the cost of waste and excess capacity in networked/shared printers could be more than $1.8 Million annually.

Personal (non-shared) printer acquisitions also become a problem in a dis-coordinated I&O environment. When users have the autonomy and budget to acquire printers (even if it is done within the constraints of a standards list), the result is a high percentage of personal printers. A survey of personal printer users across multiple corporations revealed that 90% of these users said that the reason they used a personal printer is because they print confidential or secure documents in the course of their daily work activities. The result: they get a personal printer.

If the average annual Total Cost of Ownership (TCO) of a supported personal printer is $250 and the personal printer-to-user ratio (in a 10,000 person corporation), is 1-to-4, then the cost of waste and excess capacity in personal printers could be more than $500,000 annually. Combining the dis-coordinated cost of personal printers with the dis-coordinated costs of MFPs and single-function shared printers described above, you can begin to see the impact of dis-coordination in Imaging & Output. [See the Table below]

THE 7 Cs OF IMAGING & OUTPUT

EXAMPLE: THE COST PROBLEM WITH DIS-COORDINATION

A TYPICAL 10,000 PERSON COMPANY AND HOW THEY ADDRESS THEIR REQUIREMENTS FOR HARDCOPY TECHNOLOGY

THE ACQUIRING ENTITY	THE SITUATION	TCO COST	THE WASTE (All 10,000 Users) EXCESS
A user department with 20 users	The users at the department level have the budget and the autonomy to purchase printers. many users believe they "need" a personal printer for confidentiality, and they will use their budget to purchase them. In this scenario, the user department with 20 users will purchase on average 5 personal printers [**In a coordinated environment they would only "need" one personal printer**].	$625,000	$500,000
Facilities & Operations, Sourcing, Purchasing, R/E	The Facilities & Operations department is faced with the expiration of their existing lease agreements for copiers and MFPs. They upgrade the fleet to newer devices (MFPs -without the benefit of an integrated plan with the IT department), and for simplicity, they maintain the same quantity of devices (they d not reduce the fleet size). They own the P&L for these devices even though they charge-back the user departments for their usage. [**In this scenario, they lease 2 MFPs; in a coordinated environment they would only need one**]	$8,250,000	$2,772,000
Information Technology Department	The IT department notices that the printer fleet is old and some printers are failing. They order new networked (shared) printers as well as some MFPs to help improve user productivity. They order the MFPs without the benefit of an integrated plan with Facilities& Operations. The IT department maintains the P&L for te printers and MFPs they acquire. [**In this scenario, they purchase 1 MFP and two printers; in a coordinated environment they would only needed 1 MFP and 1 printer**]	$3,500,000	$1,750,000
TOTAL EXCESS SPEND			**$5,022,000 (40%)**

Cost assumptions: Uses all TCO elements in determining costs; annual TCO for a personal printer is $250; annual TCO for a MFP is $5,500; annual TCO for a single-function shared printer is $3,500.

More Negative Effects of Dis-coordination

As stated previously, in most corporations multiple departments have purchase authority and profit & loss responsibility that overlap in the I&O environment. Dis-coordinated I&O responsibility can have additional negative results including:

- Independent purchase decisions resulting in fleet imbalance and high costs
- Multi-source budgets leading to duplication of spend and higher costs
- Inefficiency and wasted time due to duplication of efforts in pursuit of the same objectives
- Multi-source P&L responsibility for hardcopy resulting in ignored overall corporate costs
- No single department can ensure optimal deployment of technology or strategies
- Uneven buy-in to corporate initiatives such as hardcopy utility or managed models. My experience has shown that when companies attempt to implement corporate-wide balanced hardcopy models (such as managed print solutions) in a fragmented environment where users purchase and "own" their printer and MFP assets, there is often resistance to change. One reason is because only certain costs are visible to users; they might only pay for the printer hardware and supplies out of their budgets. All other cots are borne by the I/T department. When companies move to a managed print model and institute a charge-back system whereby user departments will now pay for other costs such as support, helpdesk and some I/T charges, they oftentimes execute their power

to opt-out and not participate in the program. If enough departments follow this course of action throughout the corporation, then the company will not be able to realize the full potential of their balanced/managed print solution strategy.

- The inability to effectively roll-out corporate-wide initiatives

The results of industry engagements show that dis-coordination in purchasing, ownership and P&L has resulted in up to 40% of additional cost due to excess equipment being purchased, multiple models of the same equipment being proliferated, loss discounts due to fragmented purchases and waste due to obsolescence and purchases being made without a shared vision or deployment strategy. The result, as illustrated in the table above, is that the average 10,000-person company could be wasting as much as $5,000,000 annually due to the fragmented nature of hardcopy responsibility and Profit & Loss.

Lost Opportunity to Aggregate Spend Across the Enterprise

From a financial perspective — in addition to higher costs of ownership — the lack of coordination is manifest in the money lost by not taking advantage of additional discounts that may be available to the company through the aggregation of spend across the enterprise. When companies are not coordinated across the I&O purchase and spending motions, the different purchasing entities (I/T, Sourcing, user departments) are often unaware of what each other is purchasing in other parts of the enterprise or in the different regions (for global companies). These entities are not combining purchases, tracking purchases enterprise-wide, or — as is often the case — purchasing the same technologies from different vendors. As a result, the company is not taking advantage of potential discounts that may apply based on the aggregate volume of their purchases enterprise-wide, and many are leaving millions of dollars on the table.

Dis-coordination in Enterprise-wide Imaging & Output

In addition to roles, responsibilities, P&L and office infrastructure, Coordination includes the incorporation of the various aspects of an enterprise's I&O Document-related Information Supply Chain (DISC: The flow of information from point-to-point in the business process, within and between companies, that involves paper at some point in the process. It can include business process workflow, application workflow, hardware, employee "usability" workflow and others), including remote home office-based workers, the general office, central reprographics, host/mainframe printing, commercial & external I&O and any specialty environments as described below (See *Chapter Six: Drawing Distinctions* for additional details on the various Imaging & Output environments):

- **Home-Based Office Workers**: professionals such as sales representatives, consultants, agents and others who primarily work from a home office and thus require I&O resources at the remote locations.

- **The General Office**: the traditional office workplace that houses workers in functional areas such as accounting, finance, marketing, administration, executive offices and others.

- **Central Reprographics**: also known as the central reprographics department (CRD), this entity consists of large, high output production digital copiers and MFPs that are designated for large capacity print and copy jobs.

- **Mainframe-Based Output**: print output that is produced from host-based applications such as SAP, Oracle, Levi, Ray & Shoup and others. Oftentimes, output produced in this environment is printed on green bar paper.

- **Specialty**: includes non-standard office output and can include thermal printers, label printers, impact printers, large format plotters and more. The Specialty element can also include I&O environments with atypical output requirements such as laboratories, materials printing, special sizes & substrates, MICR and others.

- **Commercial/External**: includes output production that is performed outside of the company on behalf of the company, such as brochures & marketing collateral, statements, advertising materials, packaging, regulatory reports, booklets and others. The spend for commercial/external output production is typically the largest of any entity throughout the enterprise.

An Imaging & Output strategy that does not incorporate the I&O dynamics of each of the Imaging & Output environments is too narrow in focus and will not produce the maximum impact that is achievable.

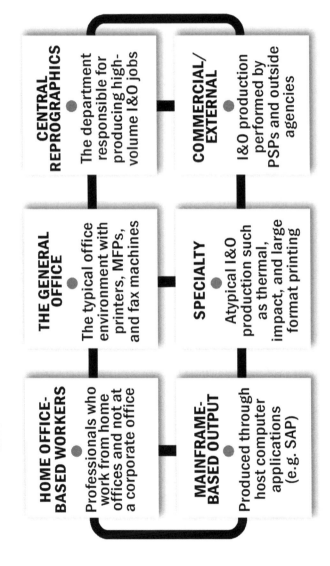

COPIOSITY

Copiosity refers to acquiring or possessing Imaging & Output technology assets large in quantity or number; abundant; plentiful. In other words, it defines the condition whereby companies — for lots of different reasons, including discoordination — acquire more hardcopy devices than are required based on the I&O requirements of the users exploiting the devices. Simply put, it defines the state of having "too many" printers, copiers/MFPs, fax machines and scanners.

In the previous chapter, I discussed the impact that a lack of Coordination can have on the superfluous quantity of hardcopy assets throughout a corporation. This is most evident in the proliferation of personal, non-shared printers in corporations.

In my experience, the majority of personal printers in corporate offices have come directly from end users, primarily when user departments have the autonomy to purchase equipment as long as the cost is below some specified dollar amount. In most of these cases, the company sets departmental budgets which allow users in the department to acquire office supplies or other "things" they need — without I/T approval — as long as that item is priced within the budget limit. This budget threshold is often in the range of $1,000 to $1,500.

When companies assess their printer fleets they typically find that the number of personal printers (non-shared printers that sit at arm's reach on a user's desk or in a user's office) throughout the environment far exceed the quantity they deem reasonable. My analysis shows that, on average, 46% of the printers in a typical company are personal printers. And in certain industries — such as the Life Sciences where the percentage of personal printers through the company is 90%

on average — the percentages are even higher.

Excess Capacity and Under-utilization

When a company is dis-coordinated from an Imaging & Output standpoint and find themselves faced with an environment characterized by too many devices, a companion reality is that this environment will consist of not only too many devices, but also with more I&O capacity than they need; excess capacity is waste and higher-cost than is necessary.

In order to operate an I&O environment that uses the devices (e.g. printers and MFPs) efficiently, the average number of pages produced on the devices per-month should fall within an output band that is generally considered to be an optimal range (from a average cost-per-page standpoint balanced against usage queuing). When corporations have "too many" devices, the average usage-per-device is far lower than is recommended, leading to excess available capacity in the devices. This device-underutilization due to excess capacity is wasteful and expensive.

An often-overlooked fact associated with having "too many" hardcopy devices is that these companies will also have too many device models (types of the same category of devices) from multiple manufacturers. When a company has too many printers and MFPs, they will inevitably have too many different models of printers and MFPs from too many manufacturers and must contend with the tedious and time-consuming task of managing the multiple print drivers. In addition, when companies have too many different types and models of hardcopy devices they must spend more time training users and supporting the variety of devices which may have different support requirements. This complexity adds to the level of inefficiency in managing such an environment.

Buying the "Boom Box"

It goes like this: Your 16-year-old nephew has a birthday coming up and you want to buy him a nice gift. You are aware that his boom-box (portable stereo system) is broken and he really loved it. You check around and you find a really fancy one that is complete with a CD player, a dual-cassette player and a radio (see below); it costs $249. Sure, there were less expensive ones available, but you wanted to get him the one with the works!

I want to draw a parallel between the boom-box-buying uncle and the copier/MFP-buying manager.

The uncle bought the boom box because he believed his nephew loved his other one and this new one was just like it, but an upgrade. Oftentimes, managers, too, simply buy the newest, upgraded version of the copiers and MFPs they currently own (or lease) just because it's newer, cheaper and "better."

The uncle bought a boom-box with a dual-cassette player on it, even though only 1% to 2% of new music is produced on that format anymore and, more importantly, no teenagers use cassettes! If the uncle had asked his nephew or took the time to look at his nephew's music collection, he would have discovered this fact. The copier/MFP-buying manager does a version of this as well. The research shows that a very small (3% to 9%) percentage of copiers/MFPs are being used to create A3 (11" x 17") size output — especially on-glass copies. Yet, the manager continues to buy all of their new MFP equipment with 11" x 17" flatbed copy capability. If the manager had surveyed the users or performed a thorough assessment he/she would have found that, indeed, there is not a requirement for a complete fleet of A3-format devices. Again, excess capacity and higher-than-necessary cost.

Self-Seeking Equilibrium

Equilibrium: [n.] a condition in which all acting influences are cancelled by others, resulting in a stable, balanced, or unchanging system. **Balance:** [v.]: to arrange, adjust, or proportion symmetrically.

The Fundamental Questions

When considering the reality that is the Imaging & Output environment of the majority of today's corporations, two fundamental questions beg to be asked:

(1) Why are there so many ("too many") printers, copiers/MFPs and fax machines in the environment?

(2) Why are the existing quantities of hardcopy devices in place — no more, no less?

I have developed a theory — "Self-seeking Equilibrium" — that attempts to provide answers to these questions. It is my belief that until managers have fact-based answers to these questions, it will be very difficult to develop effective solutions to address the situation.

My hands-on assessment, analysis & review of fifty-three (53) corporations revealed that companies that are not Coordinated and do not actively manage and continually improve their Imaging & Output environments are characterized by high costs, excess capacity, too many devices, waste and other inefficiencies. It has been my experience that managers must effectively provide answers to the aforementioned "fundamental questions" and identify the reasons in order to resolve the problems in the environment.

After an extensive analysis of assessment study data from 53 corporations where I have personally performed Imaging & Output assessments, in addition to feedback and other relevant information collected from more than 32,000 I&O users, I have come to believe that — just as "water seeks its own level" — so, too, does an organization's I&O infrastructure of equipment "seek" equilibrium and balance.

Fluid Mechanics: Why "Water Seeks Its Own Level"

Many people are familiar with the term "water seeks its own level" but fewer people know what it even means or why water seeks its own level. The concept of water seeking its own level is illustrated in the diagram below. Consider a U-shaped tube that is filled with water. The water level in both sides of the tube will eventually settle at a point that is level with each other, even though the volume and shapes of the tubes are different.

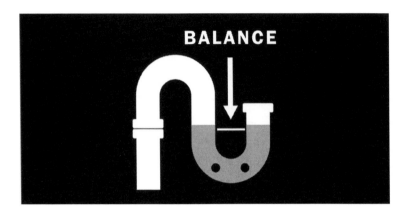

The reason water will eventually settle at equal levels in both sides of the tube is due to water pressure based on depth. Water pressure depends on depth, so only at equal depths of water will the pressure be equal. Consider again the U-shaped tube. If water is at rest where each black dot is, the pressure levels in both sides of the tube must be equal, otherwise a flow would occur from the region of higher pressure to the region of lower pressure until the pressures equalize. For this to happen, the depths below the surfaces must be equal. Or, to put it less scientifically, if left to its own devices, water will shift until it is level in both sides of the U-shape tube — equilibrium.

The point to be made through this example is that in an environment that is out of equilibrium (all acting influences are cancelled by others, resulting in a stable, balanced, or unchanging system) certain forces will come into play to ensure stability (in this case, water pressure). It is my contention that the same concept holds true when it comes to Imaging & Output hardware infrastructures that are out of balance; different forces (pressures) will come into play and have a transformative effect on the environment until the environment's I&O hardware infrastructure of printers, copiers/MFPs and fax machines achieves equilibrium and balance.

Equilibrium and Balance

It is my contention that the number of Imaging & Output devices (printers, copiers/MFPs and fax machines) in a corporation's general office environment will continue to grow until the influences of one set of resources (users and their I&O needs) are balanced against the influences of a second resource (technology infrastructure). When the I&O needs of the users are balanced against the capacity, proximity, performance, features and functions of the available I&O devices, it is at that point that equilibrium is achieved. And when equilibrium is achieved, the growth in the number of I&O devices becomes stable and unchanged.

When it comes to I&O devices, equilibrium does not automatically equate to balance. Equilibrium implies alignment between users' I&O needs and the services offered by the I&O devices in the environment. Balance assumes equilibrium PLUS the "right" quantity and types of devices; not too many so as to be wasteful and not too few so as to contribute to user workflow inefficiency.

Through my analysis, I will attempt to show that, regardless of the type of solution a company employs as a means to optimize its I&O infrastructure, the number of I&O devices in the environment will continue to increase until users' requirements for I&O resources are fulfilled.

[*Note*]: An obvious influencing resource that could counteract the influence of user I&O demand is money. If a company's executive management mandates that there will be no additional capital spending on I&O technology, then device growth will slow-to-stop. Though users can (and often do) still purchase personal printers with their own money, the number of devices that will be added through that approach would not be enough to provide equilibrium.

Balance and Efficiency: The Importance of Balance

All companies want to use resources efficiently. Efficiency, as I am using the term here, describes a condition in which assets deployed are balanced against the business requirements for those assets, at a reasonable cost. This is true regardless of which types of corporate assets are under consideration. Whether it is people, Sourcing, technological equipment, money, or fleet vehicles, corporate executives understand the importance of finding the proper balance between the deployment of these assets and the effect on *profitability*.

Imbalance between resources deployed and resources needed (based on some relevant requirement) invariably leads to higher-than-necessary costs and reduced profitability.

For example, if a company has 1,000 employees that require corporate fleet vehicles, the company will lease 1,000 fleet vehicles, not 1,200; those extra 200 vehicles would erode profitability. Or, if a company had 5,000 employees that require office space, the company will lease a facility that can comfortably accommodate 5,000 workers (ignoring growth for the moment); the company would not lease a facility that can accommodate 10,000 workers, because the excess capacity would be wasteful and negatively impact profitability.

Likewise, if a company has 10,000 users that require Personal Computers in the course of performing their jobs, the company will buy about 10,000 PCs, not 12,000; the additional 2,000 PCs would erode profitability and increase the company's waste due to excess computing capacity. However, this company would also not purchase only 8,000 PCs for the 10,000 users, because of the negative impact (and lost opportunity) that would be felt by the company due to the fact that 20% of its workforce would not have the tools necessary to do their jobs (and contribute to revenue growth) to the level that they are expected.

Companies apply the same logic for balance in the Imaging & Output environment. For instance, a company with 10,000 users will ideally provide a fleet of such I&O devices that is in balance with the requirements (both volume and features/functions/characteristics) of the 10,000 users as dictated by their use of these devices in the course of performing their jobs. Companies realize that too many devices would lead to high costs, waste and excess capacity and too few devices would lead to user inefficiency, high Document Production Costs (DPC) and ultimately, higher-than-expected total costs and reduced profitability.

Document Production Costs are a measure of efficiency and are the costs (both hard and soft) that actually go into users' production of paper output, including Total Cost of Ownership (TCO). It is TCO plus the time-related productivity costs associated with users going through the motions of sending an output job to the printer/MFP, finishing the document(s), walking to retrieve the job from the printer/MFP, taking any additional actions with the document and returning to his/her desk.

DPC = (Total Cost of Ownership) + [(Total hours required to produce & retrieve output by all users in the environment) x (Average hourly wage)]

The Elements of Equilibrium

Equilibrium is a function of user requirements in relation to the capacity, proximity, performance, features and functions of the available I&O fleet of devices.

User requirements. When it comes to Imaging & Output devices, users want enough printers (etc.) available so that they can create their output without having to wait to retrieve it. Users also want the features (e.g. color, speed), functions (e.g. scan-to, fax, private print), performance (e.g. reliability) and proximity (i.e. a short distance from their work station) to make their use of these devices productive.

- **Capacity.** The product of the quantity of devices in the environment and the output capacity of those devices, against the number of users that use the devices.
- **Proximity.** The distance between the I&O device and the users work station.
- **Features.** The different characteristics of the device, such as monochrome ink/toner, color, output speed, duty cycle, paper capacity and others.
- **Functions.** The different activities that can be performed with the device such as scan-to-e-mail, stapling, fax and copying.
- **Performance.** How reliable the device is as measured by uptime, availability, functionality and performance-to-spec.

So, companies are careful to provide equilibrium and find that balance between users' requirements for Imaging & Output resources and the technology resources that are deployed to provide balance between these two sets of resources, right? *Wrong!* ...

The reality is that most companies DO NOT effectively strike the right balance between their users' requirements for Imaging & Output and the number and types of devices deployed to support their requirements. The common result, based my analysis, is that such companies experience higher-than-necessary costs, "device creep," and user frustration.

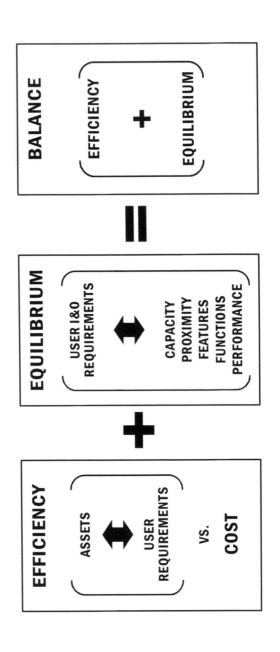

"Self-Seeking Equilibrium"

"Self-seeking Equilibrium" is a term I use to describe the organic process by which corporate offices gradually become populated with a variety of printers, copiers/MFPs and fax machines, until — assuming no mandates or effective management processes — user demand for Imaging & Output resources is met.

A review of the study data revealed that in 77% of the companies studied, the force of (imbalanced, wasteful) equilibrium was achieved in a manner typical of that illustrated by the *Fortune 500 Company* ("Company ABC") described below (whose assessment results are included among the 53 companies analyzed).

My theory of Self-seeking Equilibrium answers the aforementioned Fundamental Questions:

(1) Why are there so many printers, copiers/MFPs and fax machines in the environment?

(2) Why are the existing quantities of devices in place — no more, no less?

The Process Of Self-Seeking Equilibrium at "Company ABC"

1. In the beginning, the company opened a new office and populated the office with printers, copiers/MFPs and fax machines. The I/T or Sourcing managers placed a quantity of these devices that they estimated to be a reasonable deployment based on the number of users that would be located at the office.

2. The Imaging & Output fleet of devices in the new office was fundamentally unmanaged; there was no oversight,

no on-going fleet optimization and no continual process improvement.
3. Business units at Company ABC had the autonomy to buy capital equipment (including printers, MFPs and fax machines) without the approval of the I/T organization as long as the expenditure was below some predetermined spending limit; in this case, $1,000.
4. As more users populated the new office and as users had changing, new or unmet I&O requirements (such as color, finishing, confidential printing and proximity), users either acquired them through the I/T organization (with no needs analysis performed) or by exercising their capital expenditure option. *The key is that users had changing, new and/or unmet needs.*
5. Comfortable with the knowledge that they need just ask and they would receive a printer, users began to request printers for many reasons. In user surveys and interviews, 90% of users said the reason they "needed" a personal printer was because of confidential printing of some type.
6. This cycle of: changing/new/unmet user needs > requests for new printers > purchase of new printers continued until user I&O needs were satisfied (barring any external intervention). In some cases, that point was only achieved when, on average, each user had a device (a user-to-device ratio of 1-to-1). This process of gradually-increasing device quantities is also known as "device creep."
7. This state — where there is equilibrium between user needs and devices — was achieved without any proactive intervention by the I/T or Sourcing managers. This is an example of Self-seeking Equilibrium.

Cash-Flow

Cash-flow refers to the pattern of income and expenditures, and the resulting availability of cash; the amount of cash a company generates and uses during a given period of time.

Un-managed, dis-coordinated imaging & output environments are notoriously high cost from a Total Cost of Ownership (TCO) standpoint. Total Cost of Ownership in Imaging & Output is the true measure of what it costs to own an asset. It is the total cost of owning and making I&O assets available to users over an extended period of time. The cost categories used to determine TCO typically include: hardware, software, consumables, management (including networks) and administration, infrastructure, end-user support and service/maintenance. When you consider the total costs involved in providing office I&O services to an organization with, say, 50,000 users — at an average annual cost-per-user of even $700 — you can begin to see the magnitude of the cost. You should also be able to imagine the magnitude of the waste and excess spend to support a company of this size assuming that 50% (or pick your number) of the existing hard-copy fleet is overkill.

In most companies, I&O is treated as a service and is not a revenue generator, but is, instead, a cost of business requiring cash outflows.

TCO ELEMENT	DESCRIPTIVE INFORMATION
HARDWARE	Acquisition costs (lease, rent or purchase), annual depreciation (device age is important here), expensed cost threshold. Remember to include print servers
SOFTWARE	License costs, upgrades, customization
SUPPLIES AND CONSUMABLES	Paper, toner (percent color and mono), drum units, ink cartridges, maintenance kits
FACILITIES COSTS	Power consumption (considering power save modes), physical space (costs for space for floor-standing equipment), upgrade costs (e.g. adding power outlets, platforms, etc), phone line and port charges, purchasing costs
DOCUMENT PRODUCTION	The user workflow action (create, print, retrieve, copy, fax, return), user training, document delivery
MAINTENANCE AND SUPPORT	Preventative maintenance, break-fix service, warranty and warranty-extensions, on-site support personnel Helpdesk Costs Level-0, level-1 and level-2 helpdesk support

I do realize that some businesses actually use their I&O resources to generate revenue, but for the purpose of this book, I will focus on the typical implementation of I&O found across most corporations, and that is I&O as a service (cost or expense).

Cash-flows are often used to evaluate the state or performance of a business or project. The consideration of I&O as a negative cash-flow entity makes it interesting when it comes to building a business case for I&O initiatives and developing the cost-benefit analysis and the Return-on-Investment (ROI). Since the formula for Present Value and Net Present Value (NPV) — one of the fundamental metrics used for cost-benefit analysis — involves net cash flows (cash coming in and cash going out), it is difficult to express this metric since there is no cash flowing in and only cash being paid out to support I&O.

There are, however, a couple of approaches I have taken in the past that reflect the true financial benefits of implementing an Imaging & Output cost-reduction initiative. The one approach I will mention here is to track the Current Sate costs (cash outflows) against the Future State costs and to measure the cost differentials between the two. The cost savings between the Current and Future States can be tracked and reported; this will show the financial benefits of engaging in an I&O optimization initiative.

Course

Course is defined as the path, route, or channel along which anything moves, including information and paper.

The Document-Related Information Supply Chain (DISC)

The information supply chain is the flow of information from point-to-point in the business process, within companies and between companies. The *Document-related Information Supply Chain* (DISC) is the information supply chain that involves the creation and production of documents at some point along the process. A "document" can be defined as a written or printed paper providing information. So, the DISC is the flow of information throughout a company that at some point involves paper in the process. In 2003, I coined the term "Document-related information Supply Chain" to describe the organizational processes relate to the combination of information flow and paper capture and/or output production.

THE 7 Cs OF IMAGING & OUTPUT

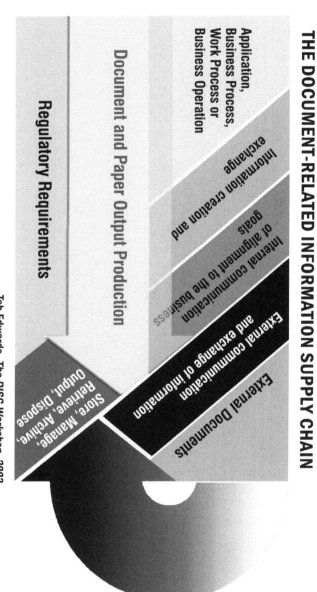

Tab Edwards, *The DISC Workshop, 2003*

A common goal in the DISC for most companies is to improve efficiency and reduce costs by creating processes that reduce or remove the number of manual interventions and the amount of paper (and paper output) involved in the processes — while maintaining compliance with relevant regulations. To that end, companies are developing and implementing company-wide document strategies and implementing new technologies. Many of these companies are developing initiatives to exploit their existing infrastructures and their existing investment on document and content management applications.

Though many managers formulate idealistic visions of how their DISC environments will be improved when they roll out document management solutions as part of a paper-based process-improvement initiative, it is not uncommon to succumb to the reality that is enterprise content and document management: one-third of all systems fail within 2 years! The typical reasons for such a high failure rate are:

- There is a rush to deploy technology without understanding the business requirements
- The solutions do not have enough impact to gain widespread acceptance and interest
- The solution that is implemented solves the wrong business problem
- Users don't want it or "need" it; it doesn't help them
- The failure to integrate with legacy applications
- Misrepresentations by the vendors selling the solutions

Even though managers consider ECM and other document solutions to provide productivity and cost benefits, managers know that they must deal with the short-term challenge of improving the existing I&O environment and reducing the inherent costs.

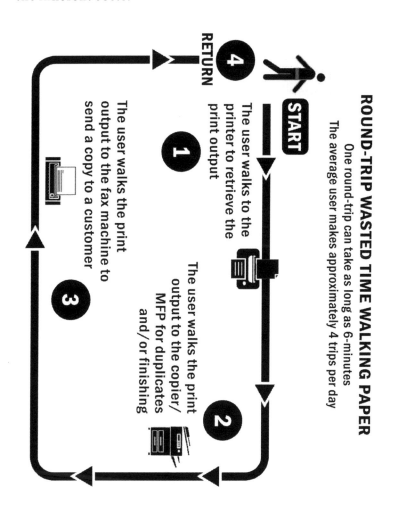

Cognizance

Cognizance is having awareness or knowledge of the cost, infrastructure, fleet and other Imaging & Output-related information about the organization.

Every company wants to reduce the cost, inefficiency and general dis-organization of its I&O environment. In order to effectively reduce Imaging & Output TCO, optimize the fleet, reduce waste, improve user-related inefficiency and reduce the amount of paper-based processes throughout departments, a company must first understand the degree of high cost and inefficiency that possibly exists. I believe one of — if not thee — most important steps involved in the I&O improvement process is to gain a thorough understanding of the current state of being (related to I&O) throughout the organization and to gain an understanding of opportunities for cost reduction and improvement.

I estimate that fewer than 25% of companies can reasonably approximate (within a 10% margin of error) the number of printers they have across the enterprise. And when you factor in the copiers, MFPs and fax machines, my estimate falls to below 5%. And to make matters more mysterious, I don't believe that many companies can offer a reasonable guesstimate of the degree of excess capacity (unnecessary, inappropriate equipment) that exists across the print, copy and fax fleets. Far too many companies lack asset tracking & management and job auditing tools that can help them gain deeper insight into their I&O environments; this information will enable them to make better, more informed decisions.

When you consider the cost of I&O, companies struggle with adequate information here, too. Many companies can tell you within a reasonable range of accuracy the amount of money they spend on the hardware, ink & toner and service — the visible costs as some call them. Unfortunately, far too many managers cannot tell you, or even approximate the amount of money they spend on Imaging & Output from a TCO standpoint. Cognizance around TCO should be one of the cornerstones of any I&O improvement initiative and strategy.

Gain More Cognizance

Once a company determines that it might be operating an I&O environment that needs to be improved upon, it is time to gain more Cognizance about the environment in order to accomplish the following:

1. Identify the factors that detract from the company's ability to accomplish its business objectives. Is the company spending too much money? Is the company wasting money? Is end-users' time being wasted on un-productive processes? Can business processes be completed more efficiently?
2. Gain an understanding of the magnitude of the problem in order to determine if the problems are worth fixing. Is the company spending, wasting, or losing enough money for you to care? Are the existing inefficient workflow processes causing the company to lose out to our competitors? Are those inefficient processes hindering the company's ability to realize more revenue?
3. Identify specific areas of the environment that can be targeted for improvement — improvements that will yield the most bang-for-the-buck.
4. Understand the elements of a strategy that will have to be developed in order to right the ship.
5. Develop a cohesive I&O strategy

Where Does the Paper Go?

In 2005, Marketing & Business Integration (MBI) conducted a study to determine what happens to all of the paper that gets produced, transferred, handled and stored (in the general office environment). The results are provided in the table below. In short, MBI found that 30% of all paper/documents that are produced in the general office place are *discarded*. They also found that an additional 10% of pages are simply *lost*. Together, approximately 40% of all of the paper that is produced in the office (printed, copied, received) is wasted.

THE PAPER TRAIL	
What ultimately happens to paper produced in the office?	
Filed/stored as paper	32.8%
Discarded	30.2%
Transferred to someone else	14.6%
Lost (70% must be re-produced)	10.0%
Converted to digital form and stored	11.3%
Other	1.10%
SOURCE: Marketing & Business Integration (MBI), LLC. Study: *Follow That Paper!*, 2005	

Without knowing such information I believe it can be more challenging to develop an I&O document strategy that satisfies the requirement for eliminating or significantly reducing the amount of paper waste and the associated costs.

Consumer

The *Consumer* is a person or organization that uses a commodity or service. It is the people that use Imaging & Output products, services and solutions in the course of performing their jobs.

The users of hardcopy technologies and Imaging & Output solutions are often the catalyst for upgrading the I&O environment and the associated hardcopy fleet of devices, however, simply replacing an old or broken device, or upgrading to a newer model often does little to satisfy the true requirements these Consumers have. Consumers (users) need tools and resources to help them do their jobs more efficiently and to allow them to be more productive. Yes, newer, more feature-rich devices can help users become more productive, however, not enough attention is paid to training and transition management initiatives to make sure that users can adequate exploit the newer technologies. Oftentimes, new MFPs will provide no user-related workflow benefits because the users have not been sufficiently educated (with reinforcement) on how the functions of the newer devices can be applied to their job functions. The result is that the company has invested in a fleet of feature and function-rich devices that are only being partially used by the users. This is another example of waste, but it is also an example of the Consumer not getting what he or she needs to do their jobs more efficiently.

Device Creep

When users experience productivity and dissatisfaction issues, or they believe their needs are not being met with the current infrastructure, they will exercise their purchase autonomy and acquire personal printers, inkjet printers, network-capable printers and even workgroup-class MFPs until they are satisfied that they have the hardcopy resources needed to do their jobs more efficiently. The result is a steady increase in the number of devices into the environment. This is what I refer to as "device creep," and device creep is expensive. I talk more about this subject in the section below.

By providing users with the features (e.g. color, speed), functions (e.g. scan-to, fax, private print), performance (e.g. reliability) and proximity (i.e. a short distance from their work station) they demand of their hardcopy fleet, users see less of a need to create "device creep" and acquire additional devices to satisfy an unmet need.

User-related Efficiency and Wealth Creation

As described earlier, calculating Document Production Cost (DPC) as opposed to simply TCO forces managers to consider the users' experiences and the productivity gains to be achieved through a successful Imaging & Output improvement initiative.

Reducing unproductive work processes can save users' time, allowing them to get more done. The added throughput (or productivity if throughput sounds too mechanical) can contribute to additional wealth-creation for the company. For example, If a company with 50,000 users makes I&O changes that result in an additional $30,000 in profit per employee, that would mean $1.5B in extra profits. And since these profits would be what economists call "rents" — additional earnings requiring no additional incremental investment in capital or labor — they would create $1.5B in new wealth (at a 10% capitalization rate). This suggests that improving user-related I&O productivity can have a financial impact as well as an employee satisfaction impact.

Deployment Models for Fleet Optimization

My years of experience analyzing, consulting with and working with corporations on the topic of Imaging & Output optimization supports the study data that almost every company that has not undergone a rigorous analysis combined with a fleet-optimization and ongoing management initiative based on the results of the analysis, is living with an environment characterized by high costs, waste and user dissatisfaction. In other words, these companies inevitably have too many devices and the "wrong" kinds of devices.

Over the last handful of years, companies have begun to take steps towards understanding their Imaging & Output environments and using the intelligence to make decisions about **fleet deployment models**, in hopes of reducing costs, eliminating waste and improving the user experience.

Fleet deployment refers to the scheme used to determine which Imaging & Output devices to purchase and where to locate these devices throughout the office place in order to provide balance between people resources and technology asset resources.

Three Common Fleet Deployment Models
(As I Define Them)

My analysis revealed that companies (and their consultants & technology representatives) generally employ three different broadly-defined flavors of deployment models as they attempt to provide balance between the Imaging & Output fleet and user demand for imaging & output resources.

1. The Spartan Model

The "Spartan" deployment model is characterized by few devices (typically 1 to 4 per floor) placed at adjacent ends of the floor (or in the middle of the floor if there is only a single device). The most common objective of companies that employ the Spartan model is to reduce the cost of Imaging & Output drastically, with less regard for user satisfaction and Document Production Costs. This deployment model is illustrated below: 120 users, 3 devices (MFPs).

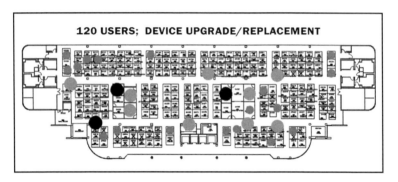

The Spartan deployment is, admittedly, a dramatic shift from the current state of nearly every company reviewed for this paper. Today, offices are filled with all varieties and significant quantities of printers, MFPs, copiers, fax machines and scanners. Under the Spartan model, this all goes away, making way for a lean fleet deployment that is reminiscent of the 2-tier, central computer deployment models of the 1980s.

My research shows that companies typically get from point (A) — the device-rich output fleets in their current states, to point (B) — the lean, Spartan deployment state — in one of two ways:

- **Corporate Mandate.** Under a corporate mandate, a company executive will mandate that all printers and departmental MFPs/copiers be removed and that all users will use the new (few) devices being installed. Although harsh, I have found this approach to be the most successful at ensuring that all existing devices are indeed removed and that the Spartan model for device deployment is established by-the-letter.

It should go without saying that user productivity is negatively impacted and user satisfaction decreases under this scenario. However, since the mandate was issued by executive-level management, users realize they have to accept the decision and adapt their document-related work habits to the realities of the Spartan deployment model for Imaging & Output.

- **Phased Device Reduction.** The theory behind this approach is that the company will implement a Spartan deployment model of a few managed MFPs strategically located in the office space. At the time these devices are installed, the company will remove a percentage (say, 5% to 50%) of the existing older printers, MFPs and fax machines from the environment and recommend that users drive their output production to one of the new larger MFPs. Over time, it is expected that users will drive all of the output to the new devices, making the current state devices obsolete. At this point (or at the point that the older devices break), the company would stop supporting the older devices and remove them from the environment.

  The financial and efficiency benefits of phased device reductions are best achieved by ensuring that the remaining non-Spartan devices are removed from the environment. In the cases I have reviewed, this was infrequently — if at all — the case. The evidence suggested that companies either removed no devices when the Spartan MFPs were installed, or they neglected to remove any devices (or prevent new devices from coming in) after the MFPs have been in place. Because these costs must be accounted for in the TCO analysis, companies in this situation failed to achieve the expected cost benefits of the Spartan deployment model.

As with most technology changes in the general office, there are pros and cons associated with the changes. Following are some of the positive and negative aspects of moving from a robust, device-heavy Imaging & Output environment to a Spartan model.

The Positive

- Lower Cost. By removing all devices and placing (for instance) three high-capacity MFPs at opposite ends of the floor (as illustrated in the diagram above), the companies that I have reviewed were able to dramatically reduce their Imaging & Output TCO in the short-term — assuming they removed the existing "older" devices from the environment.

- Ease of Administration. Having to manage and administer only a few devices per-floor is relatively easy to accomplish.

The Negative

- Reduced User Productivity. Under this model, users experienced longer document production times, a mismatch (imbalance) between the types of output-producing resources available and the functionality required and wasted time "walking paper." Walking paper is the round-trip process users typically go through in their production and retrieval of the output they produce and handle.

- User Dissatisfaction. Under the Spartan model, users consistently complained that they experienced significant retrieval time, wait times and long queues when attempting to retrieve their output from one of the few devices. Users also complained that when one of the MFPs was down (in-operational) they wasted even more time and experienced even longer queues when attempting to retrieve their output from the second MFP.

- "Device Creep" / Self-Seeking Equilibrium: Virtually all of the companies that I have studied had/have policies in place that allow users and/or departments the autonomy to purchase equipment (such as printers) without the approval of I/T, as long as the cost of such a purchase is less than some set dollar amount — often, $1,000. Users can often even acquire equipment through the normal I/T process without having to go through an executive-justification process.

  Data from the companies studied for this analysis reveals that when users under the Spartan deployment model experience the inevitable productivity and dissatisfaction issues, they exercise their purchase autonomy and acquire personal printers, inkjet printers, network-capable printers and even workgroup-class MFPs. The result is an "unauthorized," steady increase in the number of devices in the Spartan environment.

  This is what I refer to as "device creep," and device creep is expensive. Why? Because the "unauthorized" devices are not acquired under any cost-saving model and they are eventually added to the company's support agreement. Supplies and consumables are also purchased (ad-hoc, in many cases) and repair calls for these devices are placed to the company's help desk. These factors contribute to higher-than-necessary TCO.

What was disturbing to most managers that I worked with during these analyses was the fact that the costs for the "creep" devices are often not centrally budgeted or tracked, they are unmanaged and difficult to remove (as part of an efficiency initiative) since the purchasing-user considers the "unauthorized" devices to be "theirs."

- Higher-than-Average DPC. The combination of users' long, inefficient output-retrieval times combined with the higher TCO of "device creep" results in a higher-than-imagined Document Production Costs.

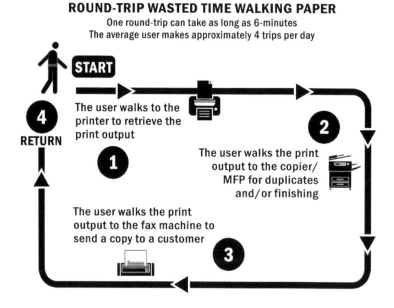

- It Misses the Point of Balance. The Spartan deployment model, by its very design, is incompatible with a balance between users (resources) and Imaging & Output technology assets (resources). It doesn't require any analytical effort to know that a user-to-device ratio of 40-to-1 (as illustrated in the diagram above) is off-balance and, therefore, highly-inefficient.

A Financial Analysis for the Spartan Deployment Model

So, how does the Imaging & Output financial model look for the companies following the Spartan deployment model? An analysis of the companies studied for this white paper reveals a cost-saving model that is less financially-attractive than was anticipated when the decision to employ the Spartan model was made. The primary reasons — as I listed above — are because companies leave older, non-Spartan devices in the environment and/or "device creep" occurs and the cost of these devices must be accounted for in the financial analysis.

The Spartan Model: A one-floor scenario (averages based on study group data and assumptions)*

CASHFLOW	YEAR 1	YEAR 2
Current State Costs	$78,480	$78,480
"Device Creep"/Self-seeking equilibrium**	$0	$11,772
Spartan Model Monthly Cost	$54,648	$54,648
Total Spartan Model Costs	$54,648	$66,420
Spartan Model Savings	$23,832	$12,060
Spartan Model Percentage Savings	30.37%	15.37

*Study group data and assumptions for this example: Assumes that every device is removed from the floor when the 3 Spartan devices were installed; 120 users on the floor; Current state devices = 60; 2 high-volume monochrome MFPs (no minimums) and 1 color MFP

(no minimums) are provided to accommodate all user demand under a Spartan deployment model; Monthly pages produced = 72,000 (average of 24,000 per new MPF); Current State monthly Imaging & Output costs (TCO) = $6,540; Spartan state monthly cost (TCO, includes applicable TCO cost factors, excluding floor space) = $4,554; Monthly cost savings of the Spartan deployment alone = $1,986 (30%); Number of "Device Creep" networked printers and MFPs added by year two = 9; Average monthly TCO of the "creep" devices = $981; Assumes no device growth beyond "device creep" (equilibrium & balance achieved).
** If all of the devices under a phase-out approach to device-reduction are not removed from the environment, the cost of the un-removed devices will increase the TCO of the device fleet (and, consequently, will reduce the cost savings).

As this example illustrates, when companies that have entered into optimization initiatives following a Spartan deployment model fail to either remove the remaining devices (after the initial removals as part of a phase-out effort) or fail to manage "device creep," the cost savings are reduced from initial expectations (from 30.37% to 15.37% in this example) due to Self-seeking Equilibrium.

A Cost-Benefit Analysis that includes Return on Investment (ROI), (Net) Present Value and Internal Rate of Return (IRR) will not yield any useful information about the attractiveness of this optimization effort as an "investment." The primary reasons are that there is typically no initial capital outlay involved in Imaging & Output optimization projects and these projects are not revenue-generating investments. That said, we can approximate the Break-even Point/Payback Period (approximately 1 month in this example) of a Spartan model Imaging & Output initiative.

2. The Fleet Upgrade/Replacement Model

The Fleet Upgrade/Fleet Replacement model is the most common improvement model employed by companies. Simply, companies will upgrade/replace their current fleet of printers and copiers/MFPs with the newer models of these devices, typically at a cost savings (since, on-par, the new devices offer the same or greater performance and functionality at a lower price).

Many managers that I interviewed find this approach to be the easiest, fastest way to reduce costs and upgrade the fleet. Though that is true, one thing that some managers fail to consider is that the old-device-to-new-device percentage cost savings doesn't translate to the overall cost savings achieved throughout the environment. The primary reason: The devices that have not been upgraded/replaced (mostly single-function printers and stand-alone fax machines) still remain in the environment and their associated TCO must be included in the overall environmental costs.

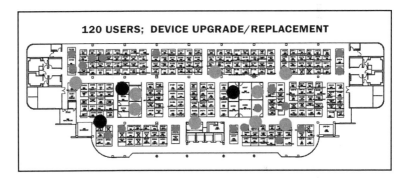

The Positive

- Lower Cost. By simply replacing a fleet of older devices (e.g. copiers and MFPs) with a fleet of the newer, lower-priced devices (acquisition cost/per-page costs), the overall costs will be reduced throughout the environment.

- Equilibrium. When companies replaced older devices one-for-one with new devices and, assuming the pre-upgraded environment succumbed to Self-seeking Equilibrium, then the upgraded devices maintained that equilibrium — not Balance, but equilibrium.

- Improved Device Functionality. Newer devices will

normally provide a more robust set of features and additional functionality than the older devices that were 3-to-7 years old. This improvement in functionality can contribute to an improvement in the user experience and efficiency.

The Negative

- Excess Capacity / Waste. By simply replacing a fleet of older devices (e.g. copiers and MFPs) with a fleet of the newer models, companies inevitably buy some percentage of devices that provide an inappropriate fit between output capacity demand and output capacity provided through the devices. In addition, some features, like 11" x 17" on-glass copy format, may not be required in the majority of the replacement device fleet (since it is estimated that only between 3% and 9% of pages produced in the general office are of this size format). By simply replacing the older model of the copier/MFP with the newer version could result in unnecessarily higher costs.

- Lower-Than-Expected Overall Cost Savings. Managers from several companies in this study estimated that — because the new replacement models were X% cheaper (e.g. 20% cheaper as illustrated in the example below) than the older models being replaced — the TCO across the overall environment would be reduced by the amount of the X% cost savings. However, when you include the TCO for the devices that were not replaced, the overall cost savings is reduced to something less than the original estimate of X%; this is illustrated in the table below.

- Imbalance. Though the device upgrade/replacement model maintained self-sought equilibrium, there was not balance. In other words, users likely had the devices they needed to do their jobs (including personal printers, etc.), but the data shows that there were still more devices than were needed to meet user requirements for I&O resources.

The Upgrade/Replacement Model: A one-floor scenario (averages based on study group data and assumptions)*

CASHFLOW	YEAR 1	YEAR 2
Current State Costs	$78,480	$78,480
Non-replaced Device Costs	$15,912	$15,912
Replacement Model Monthly Cost (save 20%)	$50,054	$50,054
Total Replacement Model Costs	$65,966	$65,966
Replacement Model Savings	$12,514	$12,514
Replacement Model Percentage Savings	15.94%	15.94%

*Study group data and assumptions for this example: Assumes 120 users on the floor; Current state devices = 60; 9 monochrome MFPs (no minimums) and 3 color MFPs (no minimums) are provided as the replacement/upgrade devices; Monthly pages produced = 72,000; Current State monthly Imaging & Output costs (TCO) = $6,540; Replacement/Upgrade state monthly cost (TCO, includes applicable TCO cost factors, excluding floor space) = $4,171; Monthly cost savings of the Spartan deployment alone = $1,043 (20%); Number of remaining non-upgraded devices = 48; Average monthly TCO of the non-upgraded devices = $1,326; Assumes no device growth beyond device upgrades (equilibrium & balance achieved).

3. The Tailored Deployment Model

The term "tailored" is synonymous with "optimized" when used in the context of Imaging & Output efficiency. That is because in a tailored fleet environment, the company will have achieved equilibrium through a quantity of devices that is neither superfluous nor Spartan, thus satisfying user demand at a desirable cost, while maintaining high levels of user satisfaction.

It is implicit in this model that companies that have optimized their environment by balancing the fleet have also instituted fleet management procedures, typically under some form of managed model. So, in addition to providing Equilibrium and Balance, companies that employed this model were most successful in preventing Self-seeking Equilibrium and the growth in I&O devices it brings.

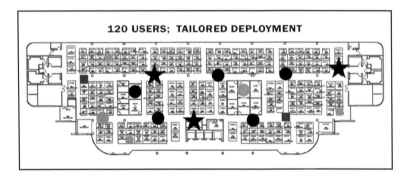

The Positive

- Lower Cost. By eliminating the excess capacity & unneeded devices and by right-sizing the fleet of equipment, the Tailored Deployment Model provided a lower cost solution that also provided Balance.

- Equilibrium & Balance. Under this model, companies were more successful at providing a fleet of I&O devices that satisfied users' I&O requirements with the fewest devices, than they were under the other deployment models. The companies not only provided an efficient, optimized environment, but also Equilibrium and Balance.

- Self-seeking Equilibrium and Device Creep Avoidance. By providing users with the features (e.g. color, speed), functions (e.g. scan-to, fax, private print), performance (e.g. reliability) and proximity (i.e. a short distance from their work station) they demand of their I&O fleet, users saw less of a need to create "device creep" and acquire additional devices to satisfy an unmet need.

The Negative

- Requires Up-front Work. In order to develop an optimized I&O infrastructure, it is necessary to perform an assessment and analysis to benchmark the current environment. This often requires data gathering, workflow modeling, fleet analysis and cost modeling. Depending on the size of the study group being analyzed, this complete process can take anywhere from 2-weeks to 60-days.

The Tailored Deployment Model: A one-floor scenario (averages based on study group data and assumptions)*

CASHFLOW	YEAR 1	YEAR 2
Current State Costs	$78,480	$78,480
Non-replaced Device Costs	$0	$0
Tailored Model Monthly Cost	$60,432	$60,432
Tailored Model Costs	$60,432	$60,432
Tailored Model Savings	$18,048	$18,048
Tailored Model Percentage Savings	23.00%	23.00%

*Study group data and assumptions for this example: Assumes 120 users on the floor; Current state devices = 60; a mix of twelve devices (single-function networked printers and networked MFPs) are provided to accommodate user I&O requirements; Monthly pages produced = 72,000; Current State monthly Imaging & Output costs (TCO) = $6,540; Tailored state monthly cost (TCO, includes applicable TCO cost factors, excluding floor space) = $5,036; Monthly cost savings of the Spartan deployment alone = $1,504 (23%); Number of remaining non-upgraded devices = 0; Average monthly TCO of the non-upgraded devices = $0; Assumes no device growth beyond tailored fleet deployment (equilibrium & balance achieved).

Your Results May Vary

My theory of "Self-seeking Equilibrium," (which is based on real world study data), suggests that the risk companies face of not matching user requirements with an optimized quantity of devices that provide the desired features, functions, performance and proximity is "device creep" and a growth in the number of I&O devices throughout the company.

I am sure some companies will have different experiences and their environments will, indeed, track more closely to the envisioned improvement model. And, sure, there are, perhaps, many reasons for the number of I&O devices in an office beyond "Self-seeking Equilibrium." That said, the data from 53 different companies suggest that Self-Seeking Equilibrium is more the norm than not.

Community

A *Community* is a social group of any size whose members reside in a specific locality, share government and often have a common cultural and historical heritage. As it relates to Imaging & Output, Community involves a company's relationship to the external environment.

Governing Bodies and Regulatory Agencies

Every company, regardless of the size or formation structure, is impacted by government regulations and must adhere to some of the many mandates that governments suggest or impose. Globally, there are hundreds of thousands of government regulations that affect companies, and compliance with these regulations can cost time and money — with no prospect of generating revenue from the compliance.

Such regulations as Sarbanes-Oxley, the Health Insurance Portability and Accountability Act (HIPAA), Gramm-Leach-Bliley (GLBA), ISO, FDA 21 CFR 11, Annex 11, Public records Office, Financial Services Authority, BSI PDXXXX, NF-Z 42, GDPdU, GoBS, AIPA, 11MEDIS-DC, Basel II Capital Accord, Electronic Evidence Act, SEC 17a-4 and others, impose financial and other resource considerations on companies just to be compliant and avoid fines and lawsuits.

Environmental Concerns

There are various state and federal laws governing the environmentally-friendly disposal of computer equipment, including hardcopy equipment and their associated consumables. Computers contain hazardous materials, including mercury, cadmium (a known carcinogen) and hexavalent chromium (shown to cause high blood pressure, iron-poor blood, liver disease and nerve and brain damage in animals). Companies that do not dispose of hardcopy equipment properly could not only be doing an environmental disservice, but could be causing harm to people and animal.

Imagine a scenario where the groundwater near a landfill becomes contaminated due to the improper disposal of computer equipment, and the contaminated water causes harm to those living nearby. If it is discovered that your company had disposed of computers at the site (identified by an asset tag on the printer or some other identifier that linked your company to the printers) the company could be subject to potentially costly criminal and civil litigation (i.e., SARA, formerly CERCLA). This could happen even if the organization had donated the equipment to a charity or paid a company to recycle it.

Energy conservation is another important topic for corporations looking to act environmentally-responsible. Over the past few years, nearly every company that I consulted with on Imaging & Output initiatives placed a strong emphasis on understanding the impact the initiative would have on power consumption.

Supplier Diversity (Minority Spend)

Though it has been a practice within the federal government for many years, recently, State and Local governments — and even many major non-government companies — are also requiring minority subcontracting. This activity, often referred to as Minority Spend, is gaining more importance as companies discover the need to establish or significantly improve Minority Spend apportionment as a way to maintain existing business and win new contracts. In fact, some companies are finding that it is sound, cost-effective business practice to integrate Minority Spend processes into their core businesses.

Competition

Although the topic of external competition does not resonate very loudly when it comes to developing Imaging & Output Strategies, it is always a good idea to consider any competitive implications of the I&O strategy that you are developing.

NINE

Managed Print Services as a Green Initiative

Over the last few years I have receives lots of requests from companies to provide "Green" workshops around Imaging & Output. These companies are interested in reducing the cost of printing while being environmentally responsible. What's interesting is the reason many of these companies give for their pursuit of a Green Imaging & Printing initiative: cost savings. In a survey of I/T professionals from over 700 companies, 65% of them said that their main motivation for going green was to save money! Unfortunately, saving money is not a green perspective. In other words, companies shouldn't go green to cut costs. The primary driver should be environmental responsibility and sustainability, with cost reduction and efficiency gains being by-products.

What Exactly is "Green" Anyway?

We hear and use the term every day, but no one really knows where it originated and why it was used as a word to describe the movement about living a more sustainable lifestyle, meaning we make responsible choices that help protect the health of our environment, our families and communities. I wasn't able to find it in the Environmental Protection Agency's (EPA) glossary of environmental terms. But whatever its origin, everyone knows that it's all about sustainability.

What's relevant to this discussion are the four perspectives (categories) of being green (Preservation, Energy, Biosphere and Global Warming) and the implications for companies as you develop Imaging & Output strategies influenced by your company's drive toward sustainability.

In order to gain an appreciation of the aforementioned Green Perspectives and why they are relevant to companies as they pursue Imaging & Output initiatives, it is worth providing you with some background on the sequence of events that brought us to this place, namely, the formation of the EPA in 1960 and the concept of "Green Chemistry."

From Earth Day to Sustainability

Many people consider Earth Day to be the catalyst for the environmental movement. What is Earth day? Earth Day is the annual U.S. celebration of the environment and a time for Americans to assess the work still needed to protect our planet. According to Senator Gaylord Nelson (Wisconsin) — the founder of Earth Day:

"The idea for Earth Day evolved over a period of seven years starting in 1962. For several years, it had been troubling me that the state of our environment was simply a non-issue in the politics of the country. Finally, in November 1962, an idea occurred to me

that was, I thought, a virtual cinch to put the environment into the political "limelight" once and for all. The idea was to persuade President Kennedy to give visibility to this issue by going on a national conservation tour. I flew to Washington to discuss the proposal with Attorney General Robert Kennedy, who liked the idea. So did the President. The President began his five-day, eleven-state conservation tour in September 1963. For many reasons the tour did not succeed in putting the issue onto the national political agenda. However, it was the germ of the idea that ultimately flowered into Earth Day."

Around this time the EPA was formed and since then, more than 100 environmental laws were passed, including the 1990 Pollution Prevention Act. In 1991 the term "Green Chemistry" was coined in the chemical industry — one of the guiltiest parties for creating chemicals and solvents that had a harmful effect on the environment. The industry suggested that "greener" solvents plus safer chemicals defined "Green Chemistry" (also referred to as "Environmentally-benign" Chemistry and "Sustainable" Chemistry).

In 1998 *The Twelve Principles of Green Chemistry* (below) were defined (Anastas, P. T.; Warner, J. C.; *Green Chemistry: Theory and Practice*, Oxford University Press: New York, 1998). Within the twelve principles lay the four Perspectives of Green):

1. **Prevention** *(a Green Principle)*
   It is better to prevent waste than to treat or clean up waste after it has been created.

2. **Atom Economy**
   Synthetic methods should be designed to maximize the incorporation of all materials used in the process into the final product.

3. **Less Hazardous Chemical Syntheses**
   wherever practicable, synthetic methods should be designed to use and generate substances that possess little or no toxicity to human health and the environment.

4. **Designing Safer Chemicals**
   Chemical products should be designed to affect their desired function while minimizing their toxicity.

5. **Safer Solvents and Auxiliaries**
   The use of auxiliary substances (e.g., solvents, separation agents, etc.) should be made unnecessary wherever possible and innocuous when used.

6. *Design for Energy Efficiency (a Green Principle)*
   Energy requirements of chemical processes should be recognized for their environmental and economic impacts and should be minimized. If possible, synthetic methods should be conducted at ambient temperature and pressure.

7. **Use of Renewable Feedstocks**
   A raw material or feedstock should be renewable rather than depleting whenever technically and economically practicable.

8. **Reduce Derivatives**
   Unnecessary derivatization (use of blocking groups, protection/deprotection, temporary modification of physical/chemical processes) should be minimized or avoided if possible, because such steps require additional reagents and can generate waste.

9. **Catalysis**
   Catalytic reagents (as selective as possible) are superior to stoichiometric reagents.

10. ***Design for Degradation*** *(a Green Principle)*
    Chemical products should be designed so that at the end of their function they break down into innocuous degradation products and do not persist in the environment.

11. ***Real-time analysis for Pollution Prevention*** *(a Green Principle)*
    Analytical methodologies need to be further developed to allow for real-time, in-process monitoring and control prior to the formation of hazardous substances.

12. **Inherently Safer Chemistry for Accident Prevention**
    Substances and the form of a substance used in a chemical process should be chosen to minimize the potential for chemical accidents, including releases, explosions and fires.

The Four Perspectives of Green

In addition to providing a summary of each of the Four Perspectives of Green, I will also provide suggestions on how companies can support the Perspectives through their Managed Print Services implementation or other Imaging & Output solutions.

PERSPECTIVE	MPS-RELATED INITIATIVES

Preservation
The preservation of nature: trees, forests, plants and natural habitats from the forces of developers.
	• Reduce consumption
	• Eliminate waste
	  – Print duplex
	  – Use recycled paper
	  – "Follow that Paper" — to eliminate waste
	  – Understand the reasons for waste and address them
	  – Web printing solution to reduce color page waste
	  – Go digital
	  – Automate paper-based processes
	  – Document Management
	  – Reduce personal printers
	  – Allocate print resources (track & monitor)
	  – Job routing based on characteristic
	  – Control printer fleet growth
	  – Print collateral on demand/JIT
	  – Purchase from environmentally-conscious companies

Energy
Concern for the depletion of our energy resources.
	• Conserve energy
	  – Reduce the number of devices
	  – Consolidate functions/devices into MFPs
	  – Upgrade to more energy-efficient devices
	  – Use Energy-Saving modes
	  – Use eco-label compliant products (e.g. Blue Angel eco-label, Canada Environmental Choice and EPA ENERGY STAR®)
	  – Manage remotely (power on/off)
	  – Purchase from environmentally-conscious companies

PERSPECTIVE	MPS-RELATED INITIATIVES

Biosphere
 	
Sustainability. Whether an item's use will be depleted or sustained naturally
	• Recycle
	  – Use recycled paper
	  – Properly dispose of old equipment & supplies
	  – Trade-in older devices
	  – Implement better supplies management

Global Warming
 	
Greenhouse gases that trap heat in the atmosphere and increases the earth's temperature
	• Reduce High Global Warming Potential (GWP) gases
	  – Use products that meet Nordic Swan and Blue Angel eco-label emission limits
	  – Teleconference to avoid CO2 emissions related to travel
	  – Limit carbon dioxide (CO2) emissions using printers with Instant-on Technology
	  – Perform a carbon footprint assessment

Developing a Green Imaging & Output Plan

A plan (or strategy) as I use the term here is a plan for accomplishing a Goal or Objective. What's the difference? A Goal is something important for your company or business unit to accomplish in order to remain viable, such as Increase Sales or Improve Customer Satisfaction or Increase Economic Profit Margin. A Goal is the intended result of a strategy. An *Objective* is a milestone that is intended to support and contribute to the attainment of a Goal. It is clearly-defined, has a measurement criteria associated with it and is intended to be accomplished by a certain date or timeframe. A simple example of an Objective that supports the Goal to *Increase Economic Profit Margin* in this case would be: *To reduce operating expenses by 12% before the close of the fiscal year on March 31, 20XX*. (For more detailed information on how to develop a robust Imaging & Output Strategy, please refer to *I&O Strategy: Imaging & Output Strategy* by Tab Edwards — Oxford Hill Press).

A simple model that I recommend for developing a Green Imaging & Output Plan is outlined as follows:

Determine what you want to accomplish

- What is the purpose of this initiative? What are you ultimately trying to accomplish by engaging in the Managed Print Services initiative? Why bother?
- Define the Goal(s) of the initiative
  - In the form of: To [action Verb] [Noun]. For example. To reduce our carbon footprint.

Identify the elements of the Green Perspectives you want to address

Refine and prioritize the list. For instance, by engaging in the Green Managed Print Services initiative, which of the Four Perspectives of Green do you want to affect for your company. The perspectives you elect to affect will impact the plan you develop and the scope of your Managed Print Services project. Whichever perspective(s) you elect to affect, they should be consistent with one of the Goals you have established for the project.

Establish where you are today vs. your goal (ideal) state

A Green Imaging & Output Assessment specifically or a standard MPS Assessment will provide the information needed to define your Current State. Based on where your company stands today, you should then determine (or at least conceptualize) where you would like for your company to be in an ideal world. Then, work backward from Utopia to something that is attainable. The difference between where you are today (the Current State) and where you want/need to be (the Goal/Ideal/Future State) is the Performance "Gap."

Define the Gap issues and the needs to close the Gap

When you define where you are today against your desired state, you can use a problem-solving method (such as difference-reduction where you list the differences between where you and where you want to be, and start reducing the differences) to list the differences between the two states. Once you have generated a comprehensive list of differences, you can then begin to resolve/fix/eliminate them to bring you closer to your desired goal state.

- Define the measurable Objectives. If, for example, one of the differences that you identify between the two states is that you have too many personal printers, one of the Objectives that you could define would be to: *Reduce the number of personal printers(which emit high Co2levels) by 50% to be completed by December 31, 20XX.*

- Define the metrics. See the example above.

Develop an action plan for achieving the Objectives

Once you have defined the Objectives and all of the differences between the current and desired states, you can then begin to develop the specific initiatives, projects, tasks that need to be accomplished in order to deliver on the specifics of the Objectives. When defining the specific hands-on tasks, I recommend that you assign an Owner for each task (the person responsible for ensuring that the task gets accomplished, not necessarily the person who will do the work), a specific completion date for each task and the resources (people, time, money) that will be needed to get the task competed.

TEN

MPS Buyers:
How to Get Started
(on the Path to MPS)

Over the years, I have worked with scores of companies on both the buying and selling sides of the MPS equation. I work with companies that are interested in developing a Managed Print Services program for sale to its customers, I work with companies that are interested in selling their existing Managed Print Services offering to prospects and customers, and I work with companies that are considering the purchase of Managed Print Services and are evaluating MPS proposals from different vendors and solution providers. Whether the company is evaluating, buying, or selling a MPS solution, they all ask me the same question: How do we get started?

Companies that are buying MPS or contemplating its purchase want to know how they begin the planning & acquisition process in order to make the process flow smoothly and effectively. Companies that are selling MPS want to know to become more effective at reaching their customers and gaining agreement from their customers to implement their MPS solution offering. And companies that are in the early stages of developing a MPS offering and rolling it out want to know how best to get started. Since these questions exist with nearly all of the companies I have ever worked with — across the globe — I assume that you, the reader, may have the same or similar questions. A comprehensive answer to these questions could fill an entire book, so therefore I have provided a concise outline of best-practices that should, at a minimum, help you in your MPS endeavors.

...

Start by developing an Imaging & Output Strategy or at least a clear articulation of your interests and intentions with improving your Imaging & Output environment. It may turn out that Managed Print Services is not the only option available to you as a means of accomplishing the objectives that you establish for your Imaging & Output improvement initiative. I highly recommend asking yourselves the following questions and *honestly* answering them based on your current situation. Your answers to these purposefully-worded questions should give you better insight into the necessity to act.

1. ***What are you trying to accomplish?*** What are you hoping to accomplish by engaging in an initiative around Imaging & Output? What is the purpose of this project? What business Objective are you hoping to accomplish or contribute by potentially implementing a MPS solution? These questions will serve as the guiding principles for your I&O improvement initiative.

2. **Why change anything?** Based on your business-related answer to this question it may turn out that you *don't* need to change anything major if you conclude that a change will not likely yield any significant improvement or get you any closer to accomplishing your Objective for the project. On the other hand, you might discover that if you don't change anything the result could be detrimental to your business.

3. **What's happened?** Has something *happened* that warrants a change? Has there been a *compelling event* (such as a flood, a fire, a department-relocation, a hiring spree, lease expirations, a mandate by the CEO that you *will* reduce costs) that means you have no choice but to buy something? Or maybe nothing has happened. And if that is the case, then you should re-evaluate whether this is a project that is "nice to have" or if it is a project that must happen.

4. **What's broken?** Why are you investigating MPS, is a process broken? Does something within the DISC not work? Is *nothing* broken? And if nothing is broken, then what are you trying to fix with a MPS implementation?

5. **How do you know it's broken?** If something *is* broken, then how do you *know* it's broken? Do you *think* it's broken, do you *know* it's broken and can you *prove* it's broken? If you only *think* something is broken, then there is an opportunity for you to do some digging (an Assessment?) to validate your assumption.

6. **Why do you "need" to buy something "now"?** If you decide that there is no *reason* for you to buy or invest in anything *now* (in the immediate-term or short-term), then ask yourself why you are investigating MPS? Possibly for a near or long-term initiative? No reason? This is surely a question

that you should answer honestly.

7. *What happens if you do nothing?* This is the key question. If you maintain the status-quo and expect nothing negative or inefficient to happen, then do you *really* need to engage in any improvement initiatives? On the other hand, you may determine that if you do *nothing*, you will not be able to accomplish your business Objective or worse still you'd be fired, then you probably do need to do *something* — and quickly!

Determine the Opportunity for Cost Reduction

Over the years I have come up with a quick & dirty approach for easily approximating a company's potential cost savings by implementing a managed Print Services solution; and believe it or not, it works reasonably well for such a simple, fast assumptive model. My research from 12-years of performing assessments and gathering assessment data shows that the average amount of money a company (across all industries combined) spends per-user-per-year in Total Cost of Ownership on printing, copying and faxing in the General Office is within a range of $540 on the low-end to $1,400 on the higher-end. This does not include such ancillary "costs" as document management costs, end-user time, or other similar costs.

Here's the quick & dirty approach for estimating your company's opportunity for cost reduction:

A. Determine the number of users in the company

B. Assume an average Imaging & Output cost-per-user per year of between $700 and $800 (TCO)

C. Multiply A by B and calculate the estimated annual spend

for Imaging & Output in your company

D. Assume a 23% cost savings from the result in step C; this is your company's estimated potential cost savings by implementing a MPS solution. Why 23%? It is a widely-used, stated average MPS savings figure that has been determined by major vendors. Based on my experience, I can attest to the fact that it is a reasonable number to use.

E. Decide: If this assumption is correct or even close, are the potential savings worth pursuing?

Here's an example:

Suppose yours is a 5,000 person company and you decide to use $800 to represent the annual TCO for general office printing, copying and faxing per-user per-year. Doing the math you would arrive at an estimated annual hard-dollar TCO cost of $4,000,000 for print, copy and fax. Assuming you could achieve the average stated MPS cost savings of 23%, then your company could save roughly $1M through Managed print Services.

Start with a Strategy

For any business initiative that supports a stated goal, the development of a well-considered plan is always recommended. That is true for Managed Print Services, too. If a company has defined a set of goals or objectives that it is trying to accomplish for the betterment of the business and the company has determined that MPS is one vehicle for helping the company accomplish its goals, then a strategy should be developed which clearly defines how their pursuit and ultimate implement of a MPS solution will directly contribute to the goal(s).

Why develop an Imaging & Output strategy? It's simple: Because you've got problems you need to fix! (Okay, maybe that's a little harsh). But really ... if your company has problems it needs to fix, improvements it wants to make, or any objective it wants accomplish, then a strategy will define how you get there and what's needed to get you there. If your aim is to maximize the benefits of your investment in hardcopy technologies, to fully exploit your I&O resources, or to execute on the objectives that you have established for your business, the development of an I&O Strategy is a proven approach for doing so.

From an executive's perspective a strategy does more than serve as a roadmap for fixing problems — because, let's face it, not all strategies deliver the expected results. A well-considered strategy helps to align an organization's resources (which is important for *fragmented* I&O environments) and determine the business(es) or projects the company will engage in. But the primary use of a strategy related to Imaging & Output is to help define how a company can get from its inefficient Current State to a more efficient Desired State. In other words, it defines the process a company will likely go through as it gradually progresses from inefficient state to efficient state, or as I refer to it, traversing *The 5 Stages of Imaging & Output Evolution.*

As described below in the next section on *The 5 Stages of Imaging & Output Evolution* (also detailed in the book *I&O Strategy: Imaging & Output*, Tab Edwards, Oxford Hill Press), in order to make *Progress* from the *Freewheeling* (inefficient) state (stage 1) towards Efficiency (stage 5), companies must perform some type of problem solving methodology to take the company from point-A to point-B; in other words, a strategy must be put in place to determine where you want to be, how you plan to get there and what resources will be required to deliver against the strategy. Fundamentally, executives develop strategies for their companies because they want to be in a situation they are currently *not*, and they need to determine how to get there. For example, a company may want to operate its general office print, copy, fax, scan and Document-related Information Supply Chain at an average annual total cost per user of $700, where today they operate at a cost of $1,200 per-user. In this case, an Imaging & Output strategy would serve as a roadmap that defines how the company will get from $1,200 per-user down to $700 per user and what is needed to get there. And when it comes to Imaging & Output, my definition of an Imaging & Output Strategy is simply: A plan that companies establish to traverse *The 5 Stages of Imaging & Output Evolution*, from "free-wheeling" towards "efficiency".

THE 5 STAGES OF IMAGING & OUTPUT EVOLUTION

Over the past 12 years, I have observed and documented the commonalities between companies that take the initiative to make improvements in their I&O environments and the courses along which these companies have traveled to arrive at a stage of improvement where the cost levels, operational improvements and overall efficiency (as measured by time, lack of complexity and money) are consistent with their stated objectives. I have observed a wide range of conditions across these companies ranging from companies that have never instituted guidelines, procedures, or controls, to companies that have instituted strict cost-reducing directives from the CEO as a means to control the costs of Imaging & Output.

An analysis of this data revealed that, at some point in

their existence — from uncontrolled-chaos through their evolution to efficiency — each of the companies analyzed could be characterized at some point in the process as existing in one of five stages of development: *Freewheeling, Understanding, Order, Progress* and *Efficiency*. And depending on where a company falls within my 5 *Stages of Imaging & Output Evolution*, it will give an indication of what the company needs to do next on the path to I&O efficiency.

This begs the question: are the stages sequential? For instance, does a company that is characterized as being in the Understanding stage have to first bring Order to the environment before it can make Progress and ultimately, become efficient? I believe the answer is, yes — with one exception. A company in the Freewheeling stage of I&O evolution must gain a level of Understanding of the conditions in the environment before it can make Progress. However, the company in the Freewheeling stage does not have to move through the Understanding stage before it can start to bring Order to the environment; Order and Understanding can begin simultaneously or they can overlap.

From that point, however, the remaining stages are sequential; Understanding and Order must be established before true Progress can be made, and Progress must be made before a company can become Efficient. [See the diagram below]

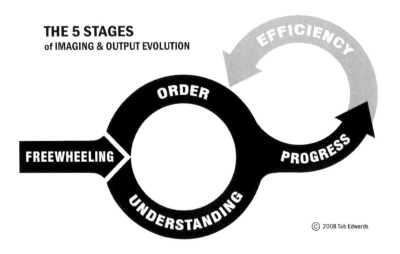

THE 5 STAGES of IMAGING & OUTPUT EVOLUTION

© 2008 Tab Edwards

Why do I believe that companies (must) follow the sequence I have defined as the 5 Stages of Imaging 7 Output Evolution in order to achieve any significant level of sustainable efficiency? Because far too many companies that were operating under Freewheeling conditions tried to make the leap to Efficiency without passing through the stages of Order, Understanding and Progress, only to find themselves compelled to spend time in these stages. It is typically not a case of, " ... well, we're struggling to make any efficiency gains so let's go to stage 2 and try to impose some Order in the I&O environment before go any further down the path we're going." Unfortunately, it is not a choice that companies make as to which stage of the evolution process they want to be in. Instead, it is a set of acting influences (see "Self-seeking Equilibrium" in chapter 3) that affect a company's Imaging & Output position, and this position will determine where a company exists along the Imaging & Output evolution spectrum. And depending on which stage your company your company exists in, that will determine the course of action you must take to move to the next stage of progression towards the objective of delivering an efficient Imaging & Output environment.

THE 5 STAGES

1. Freewheeling

The Freewheeling stage is characterized by an Imaging & Output environment that has traditionally been unmanaged, unmeasured and dis-coordinated. It is free of restraints or rules in Coordination, organization, methods, or procedure. In the Freewheeling stage, user departments have purchase autonomy within a certain dollar amount under which they can purchase hardcopy technology. Managers of environments in the Freewheeling stage will typically order and install printers, MFPs and other I&O solutions at the request of the user departments (Consumers), as long as there is funding.

Consumers in the Freewheeling stage have more Imaging & Output resources available to them than they need; in some cases, on average, each user has a printer, for instance. However, as is often the case in a Freewheeling environment, though Consumers have "too many" hardcopy devices, they may not necessarily have the "right" types of devices they need to make performing their jobs (from an I&O consumption standpoint) more productive. That is one reason why Consumers will make the request for (and receive) *another* hardcopy device that they believe has the features or functions they "need" in order to make the performance of their jobs more productive. The result is yet additional hardcopy devices in the environment, exacerbating the already-inefficient Freewheeling environment.

2. Understanding

The Understanding stage is characterized as having knowledge of or familiarity with particular aspects of the Imaging & Output environment, whether financial, operational, technological, or user-related. It assumes managers have the know-how and skill necessary in dealing with or handling I&O-related issues.

Basically, the Understanding stage is a test of the level of information and insight a company has about its I&O environment, including information about the hardcopy technologies, 3rd party relationships, applications, user-related productivity issues, cost information, usage information and general operational issues.

Action: To gain an Understanding of the Imaging & Output environment, companies must determine the questions that need to be answered and perform the investigation and analysis in order to answer them.

3. Order

The Order stage of Imaging & Output evolution is characterized by an environment that is operating properly, satisfactorily and/or under the guidance of established practices. When Order is imposed, the I&O environment seems to operate under a methodical or harmonious arrangement.

The concept of Order is akin to the state of being that an inner-city Patterson, New Jersey high-school principle named Joe Clark brought to an underperforming school when he became its principle. In this now famous story (which was made into a 1989 film entitled "Lean on Me" starring actor, Morgan Freeman), Joe Clark knew that he could not turn the school around until he first understood the lay-of-the-land and tackled the issue of drugs and students' expectations.

Upon his arrival, students were wanton in their behavior and the environment was chaotic. Eventually, after dealing with the "problem" students and engaging parents in the education process, Joe Clark was able to establish a sense of Order in the environment so that teachers could teach and students could learn. After Order had been established in the school, principle Clark was able to succeed at improving the overall performance of the students at the school, in spite of having parents, teachers and politicians against him.

The point of this anecdote is that, until Order is established throughout the I&O environment, companies will not be able to effectively make Progress toward the objective of Efficiency.

Action: To establish Order, companies must identify the areas of high cost and inefficiency, determine the reasons for their existence and establish standards, rules, guidelines, procedures, penalties and tracking/management to ensure adherence for improvement.

4. Progress

The Progress stage is characterized as a movement toward a goal or to a further or higher stage. It involves growth or development and continuous improvement. In general, Progress involves general advancement toward some goal or objective.

A simple example of how Progress emanates from Understanding and Order is the challenge of weight-loss. Millions of people struggle with weight-gain, something that prompts so many of us to make New Year's resolutions to lose excess weight. In order for us to determine the amount of weight we should strive to lose in addition to the likely causes of our weight-gain, we first must understand how much we weigh today, how much weight we have gained since we were at our "ideal" weight, what Freewheeling habits we engaged in that have added the extra pounds and what corrective actions we can take to help us lose some of the weight.

Once we make the decision to stop the Freewheeling practices that have contributed to our weight-gain, we then seek to understand our current position and identify areas where we can make improvements. Once we have come to an Understanding of the issues, we can then take corrective actions by establishing Order in our eating, exercise and healthy lifestyle habits. Once we have begun taking this corrective action, we will soon begin to make Progress toward our goal of losing enough weight to fit into our prom gowns!

Action: To make Progress towards efficiency, companies must perform some type of gap-analysis or problem solving methodology to take the company from point-A to point-B; in other words, a strategy must be put in place to determine where you want to be, how you plan to get there and what resources will be required to deliver against the strategy.

Gap Analysis

After the company has gained an Understanding of the Current State of its Imaging & Output environment, developed an Understanding of the areas of high cost and inefficiency and their causes and determined the way they want "to be" over the next 12-to-18 months (their desired state of an efficient environment; the goals and objectives), they can paint the picture of where they are today, where they want to be, the differences between the two and develop strategies to get to the desired state.

THE RELATIONSHIP BETWEEN UNDERSTANDING, STRATEGY, PROGRESS, AND EFFICIENCY

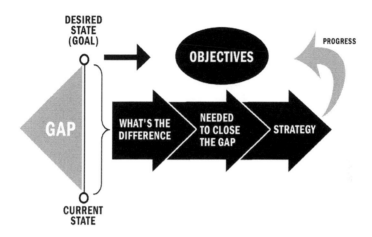

Problem-Solving Methodology

Progress is about solving problems to get you closer to your desired state (goals and objectives). When approaching a problem of the magnitude of those that confront us in the Imaging & Output arena, I recommend investigating the use of a problem solving methodology to provide structure around the process of closing the gap between the Current state and the Desired State. Specifically, I recommend the Difference-Reduction Method of problem solving.

The Difference-Reduction Method of problem solving involves Understanding where you are today, determining where you want to be in the future and identifying the differences and similarities between the two states. Once you have accomplished this, you can then put plans in place to resolve the differences between the two states until you have reached your desired (goal) state. This, of course, implies the need for a way of evaluating the differences between the two states.

Say, for instance, you conducted an assessment of your I&O environment and discovered that your company has 10,000 users, 4,000 printers and 1,000 analog copiers. Your cost analysis showed that the TCO for this hardware infrastructure is $8M annually. After you have identified the problems in the Current State, you then design a Future State environment that is clear of the issues and problems plaguing the Current State. Your Future State still has 10,000 users, but it only has 500 printers and 500 MFPs at an estimated annual TCO of $5M.

Your next task would be to determine the best way to get from the 5,000 hardcopy devices in the Current State to the 1,000 devices in the Future State, all while taking into consideration the 10,000 Consumers and their user experience. Each time you are able to remove a printer and an analog copier — while giving the Consumers more Imaging & Output resources to help them be more productive in their use of the technologies — you are reducing the differences between the two states and making Progress towards your ultimate goal of Efficiency. And the more changes you can make such that the Current State starts to look more like the Desired State, the closer you become to accomplishing your objectives.

5. Efficiency

The Efficiency stage of Imaging & Output evolution is characterized by the accomplishment of or ability to accomplish an I&O-related activity with a minimum expenditure of time, money and effort — the key measures of Imaging & Output Efficiency. When a company has improved its performance and is performing at a greater efficiency level (which can be measured in many ways depending on how the company sees fit to measure it), the high-performing organization tends to contribute to the advantages to the overall corporation.

In most cases, some measure of efficiency — whether it's cost, productivity, or lack of complexity — will be an established goal or objective of companies trying to make improvements in Imaging & Output. Accomplishing a series of objectives will lead to the successful satisfaction of an associated goal, and various strategies should be developed to accomplish the objectives that will lead to the satisfaction of an associated goal (a more comprehensive overview of goals, objectives and strategies will follow in a subsequent chapter of this book). It is important to note that once a company has successfully accomplished its objective of Efficiency, the work is not done. You must maintain a degree of Understanding of the I&O environment and continually check to ensure that Order is being maintained. You must also continually improve by making incremental improvements (Progress) in order to maintain optimal Efficiency over time.

The key take-away from this discussion is that the strategy a company develops is the vehicle by which the company will arrive at their Desired State (the goal). Therefore the development of an effective Imaging & Output strategy is paramount if a company wants to make improvements in its I&O environment.

A Value-Driven Strategy

A good strategy should be guided by a vision of how your company, division or department will create *value*. Therefore, the first step in developing a strategy (unless you are improving on an existing one) should be to define the *purpose* of the organization, which is dictated by the *value* an organization delivers. One way to begin thinking about your organization's unique value is to consider the following:

If your organization went away today (in other words, your organization no longer exists), who would care and why? Now, imagine your organization still in existence (I'm sure it's not hard to imagine that scenario) and ask yourself: What is the difference between the company *without* my organization and the company *with* my organization? The answers to these questions should give you an idea of what your *value* is and, in turn, what your organization *exists for*; the strategy that you develop should support the things that your organization exists to do.

Developing an Imaging & Output strategy can be a daunting task depending on the business objectives being addressed, the level of complexity in the organization and the scale and scope of the problems which the strategy is intended to resolve. For instance, I once facilitated an I&O strategy development project for a *Fortune 15* corporation that took — what seemed like — forever to complete. In another case, I facilitated the I&O strategy development project for a *Fortune 20* corporation that only took two weeks. The different scale, scope, complexity and "readiness" of the two corporations dramatically influenced the duration of the strategy development initiatives between these two similar corporations. But the encouraging part is that in the end, an effective strategy was developed that led to the accomplishment of a defined set of business objectives. So in the case of the *Fortune 15* corporation, though the strategy development process was challenging, the resulting benefits far overshadowed the obstacles we had to overcome to develop it.

A Simplified Model for Developing An Imaging & Output Strategy

I'll be artless here: when it comes to the development of a practical Imaging & Output strategy, companies want a simple, easy-to-develop-and-execute, practical (as opposed to theoretical or academic), plan for "improving" their current I&O environment in the short-term; this is a reality that I have come to acknowledge. And for such companies, I propose a straight-forward, 10-step approach for doing so (excerpted from *I&O Strategy: Imaging & Output*):

1. Alignment Check

Assure that every team member (especially I/T and Facilities/Purchasing/Real-Estate) is on the same page and that there is coordination between and among the different business units that have some ownership and, profit & loss responsibility for parts of the DISC, such as hardware, supplies, helpdesk, CRD/mailroom, documents, fax and record management.

2. Inventory the I&O fleet and monthly page volumes (an Assessment)

3. Review the Existing DISC Processes

What are the processes users go through in the course of doing their jobs that involve the use of hardcopy technologies at some point the process?

4. Calculate the Total Cost of Ownership (TCO) and the Document

Production Costs *(DPC)*

TCO is the total cost of acquiring an asset, owning the asset and making the asset available to users over an extended period of time. DPC are the true reflection of what it costs to product documents/paper output and it includes TCO plus user-related costs calculated by multiplying an average user's hourly-wage by the amount of time that is spent by the users to produce a document.

5. Define the Existing Issues and Problems

Once you have become aligned on the project and have gained some level of "understanding" about the current state of the environment, you can begin to identify the areas of excess capacity, waste, "high costs," and inefficiency.

6. Refine the List of Problems

Narrow the list to those things that support your objectives for the project, have the greatest impact and can be accomplished in a reasonable amount of time.

7. Conceptualize the "Improved" Future State Environment

Or the *Desired State* environment

8. Confirm and Validate the Objectives of the Improvement Initiative

You want to determine what it is that you need to accomplish to fulfill the purpose of *the project*. What is the measure (objective)? When must it be accomplished by?

9. Determine the Actions necessary to Accomplish the Objectives

Some helpful Planning Questions are:

- What are we trying to accomplish?
- What do we want to change (do differently)?

- How do we know what to change (or buy)?
- How will we know that a change is an improvement?
- What changes can we make that will result in an improvement?

10. Optional: *Build the Business Case*

I recommend building a comprehensive business case as a means of "selling" the ultimate improvement project to gain go-ahead approval and funding — assuming those things are needed. A general approach for Building a Business Case includes the following:

- Define the Project Purpose
- Establish a Benchmark (Current State)
- Design a Recommended Future State
- Conduct a Technology Assessment
- Identify the Expected Changes brought on by MPS
- Perform a TCO Analysis
- Perform a Cost/Benefit Analysis (focusing on ROI and payback period)
- Develop a Project Timeline
- Perform a Risk Assessment
- Develop a method for Results Validation

ELEVEN

MPS Sellers: How to Get Started
(on the Path to MPS)

Working with MPS vendors and solution providers big or small, domestically or abroad, a pattern emerges that reveals the major challenges and issues these companies experience when trying to establish, develop, gain traction with, sell, or grow their Managed Print Services practices. Whether yours is an established *Fortune* 200 Corporation trying to improve your sales team's performance in selling the solution, or if you are a channel partner trying to understand why your MPS program is not "working" as you hoped, chances are you are being affected by one of the following issues:

1. Where and how to start (*"Start–Stop–Learn–Start–Repeat"*)
2. How to structure their MPS solution offering and Partnerships
3. Sales Rep Skills: Transition from transactional to solution sellers

4. Properly compensating the "box" sales reps; adapting the compensation plan
5. Failure to implement a first-rate Assessment methodology
6. Failure to commit to offering Managed Print Services or to understand that it is an investment.

1. **Where and how to start ("Start–Stop–Learn–Start–Repeat")**

Companies that are relatively new to offering MPS (offering it for 3-years or less) often go through a process that I have dubbed the "Start-Stop-Learn-Start-Repeat" process. These companies will *Start* offering a "MPS" solution, get a live prospect or customer, they *Stop* offering the "MPS" solution because they have encountered things that they have not anticipated. Then, after taking their bruises and *Learning* from this experience they will *Start* offering "MPS" again, and the process *Repeats* itself. Some suggestions for ending this cycle include the following.

Define your purpose for offering MPS

Determine why you want to offer Managed Print Services. Is it to increase profitability or simply to increase sales? Is it to capture market share or to put your major competitor out of business? Is it to find new revenue sources or to provide your sales reps with a growth & development opportunity? Determining the reasons for your pursuit of MPS will help you to develop an appropriate strategic plan to help you accomplish them.

Develop a strategy to get there (start small then grow)

Far too often companies are so anxious to enter the MPS arena that they do so far too optimistically and aggressive. I recommend developing a MPS program strategy with a focus on getting into the business less aggressively. Start small, learn how to do it right, then gradually grow the scope and scale of your program.

Do a quick-n-dirty GAP analysis (including the type of solution you <u>want</u> to offer)

Determine what capabilities you have today and the type of MPS solution (simple/basic or robust) and identify the differences.

Take an inventory of people, hardware, software, capabilities

What parts of a MPS solution do you sell today? What MPS skill sets do you have in-house today?

Determine what else is needed to close the Gap

When you have identified the list of Gap items between the MPS capabilities you have available today and the MPS capabilities needed to get to your desired state, determine how to eliminate the Gap items and to get the elements to get you closer to where you need to be.

Figure out how to get those things

Use partners, relationships, other available sources, or acquire the things you need to get you to your desired MPS state.

2. **How to structure their MPS solution offering and partnerships**

Based on the possible MPS Solution Elements that can be included in a MPS offering (see the "Managed Print Services Solution Elements Table" below)

- Determine the things you *want* to offer

- Determine the things you *can* offer
- Determine how to get the things you *can't* offer

Define the Sale-to-Support Process (see the *Sample Sale-to-Implementation Process* diagram below as an example)

MANAGED PRINT SERVICES: SOLUTION ELEMENTS (OPTIONS)

PRODUCTS	SERVICES	SERVICES (CONTINUED)
☐ **HARDWARE**	☐ **SITE ASSESSMENT**	☐ **USAGE TRACKING**
☐ Printers	☐ Assessment Tools (TCO, Baselining, etc.)	☐ For charge-backs
☐ MFPs	☐ Methodology	☐ For optimization
☐ Copiers	☐ Statement of Work	☐ **ON-LINE CUSTOMER SERVICE POPRTAL**
☐ Fax Machines	☐ Fee-based service	☐ Service Requests
☐ Manage Existing Hardware	☐ **ASSESSMENT SOLUTION/FLEET DESIGN**	☐ Account Information
☐ **SUPPLIES & CONSUMABLES**	☐ Pricing Tool	☐ Consumables Ordering
☐ Ink	☐ Design Methodology	☐ **SUPPLIES REPLENISHMENT**
☐ Toner	☐ **HARDWARE SUPPORT**	☐ Supplies ordering process
☐ Paper	☐ Remote Diagnostics	☐ Consumables
☐ Preventative Maintenance Kits	☐ On-site Support	☐ Delivery (dock, desk-side, etc.)
☐ Drums, Fusers, Rollers, Wipers	☐ Multi-vendor Support	☐ Toner installation
☐ Other	☐ National	☐ Supplies management
☐ **SOFTWARE**	☐ Regional	☐ **ACCOUNT MANNAGEMENT SERVICES**
☐ Scan-to	☐ **PHONE SUPPORT**	☐ Contract/Engagement Manager
☐ Usage Tracking	☐ Link to Helpdesk	☐ Usage Tracking & Reporting
☐ LAN-Fax	☐ **PREVENTATIVE MAINTENANCE**	☐ **EDUCATION & TRAINING**
☐ Fleet Monitoring/Management	☐ Maintenance Kit Replacement	☐ On-site training
☐ Other	☐ **DEPLOYMENT PROJECT MANAGEMENT**	☐ Web-based training
ADMINISTRATION	☐ **DISCOVERY & DESIGN (PRE-INSTALL)**	☐ Job-aids
☐ **CONTRACTS & AGREEMENTS**	☐ **INSTALLATION**	☐ Documentation
☐ Master Services Agreement	☐ Un-box	☐ **TRANSITION MANAGEMENT**
☐ Custom Services agreement	☐ Inventory & Track	☐ Training & Communication Plan
☐ Statement of Work	☐ Set up	☐ User familiarization (with H/W)
☐ **BILLING**	☐ Connection (A/C and Network)	☐ **ON-GOING FLEET OPTIMIZATION**
☐ Single Invoice	☐ Configuration (IP, N/W, Subnet, gateway)	☐ Fleet downsize
☐ Electronic Invoicing	☐ Testing	☐ Fleet/Device upgrade
☐ Invoice-by-Department	☐ Driver Provision & Services	☐ **SECURITY**
☐ Integration with in-house apps.	☐ **DE-INSTALLATION**	☐ Device
☐ **PAYMENT OPTIONS**	☐ Packaging & Re-boxing	☐ Network
☐ Base + Click Charge	☐ **EQUIPMENT DISPOSITION & RECYCLING**	☐ Document
☐ Pure Per-Page Pricing	☐ Proper disposal of old equipment	☐ Access
☐ Pre-paid Pages + Overages	☐ **REMOTE FLEET MONITORING & MANAGEMENT**	
☐ Level Estimated Payments	☐ Monitoring Tool	
☐ Other	☐ Device re-location	
☐ **CHANGE MANAGEMENT**		
☐ Install, Move, Add, Change		

MPS SELLERS: HOW TO GET STARTED

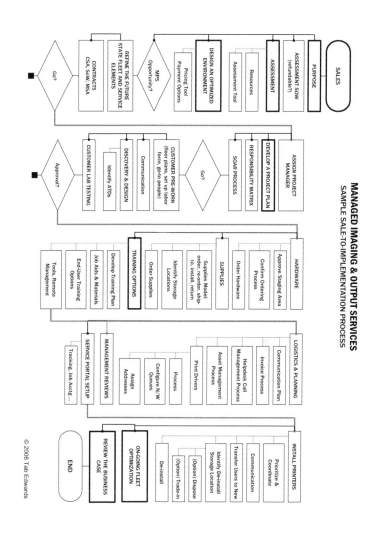

Map and define each step that is required from the sale of a MPS deal through the implementation and on-going support. Which things are you able to do today? Which things will you need to find a solution for?

Determine your Resource Gap in the Sale-to-Support Process (have vs. need)

Include administrative requirements, tools and processes (contracts, partner relationship agreements, pricing, a pilot process, etc.) How will you get what you don't have?

3. **Sales Rep Skills: Transition from transactional to solution sellers**

This is one of — if not thee — major challenges facing MPS solution providers big and small, namely, improving the effectiveness of their sales force at selling services in general and Managed Print Services in particular. Most sales reps in the Imaging & Output space started out as transactional box sellers. Many of these sales reps are from the old-school and some will never successfully transition to the more complex solution sale that is MPS. Others, however, have the capacity to sell services effectively, but getting them to that point is often a challenge. Here are some recommendations:

- Train them on Solution Selling and Consultative Selling
- Train them on the Stages to the MPS Sale
- Presentation Skills are important, so train them on effectively delivering services-oriented presentations
- Educate them on Building the Business Case (for large deals)
- Get them Certified for MPS Proficiency

    The Water Training Institute — a MPS consultancy and training organization — offers MPS sales certification for sellers that are serious about becoming qualified MPS sales professionals. Many managers who wanted to gauge and improve the sales quality level and effectiveness of their MPS sellers have found significant value in the program.

- Hire MPS Specialists or experienced sellers
- Investigate "Sales Transitioning" Support

    Some organizations offer services whereby they will sell a company's MPS solution directly for the company or help a company's sales team sell and close MPS deals while training them on the process along the way. This can be an attractive option for companies that want to test the MPS waters before going in full-throttle.
- Ensure they understand the importance of transitioning

    For many companies, transitioning to selling printer-related services and MPS is not an option. Therefore, companies must make it clear that the company is moving in that direction in order to remain competitive and viable, and that everyone needs to get on-board.

4. **Properly compensating the "box" sales reps; adapting the compensation plan**

Companies should develop a compensation plan that rewards the desired behavior of the sales team — namely, supporting and selling MPS. Managers must make it "worth it" for their sales teams to leave their comfort zone of selling "boxes" and focus their energies on growing the MPS business.

Compensation Considerations

The way you decide to compensate your sales team will have a significant impact on how successful your Print Management/ Managed Print Services (MPS) program will ultimately become. For instance, if a company compensates its sales reps based on hardware unit volume or the associated revenue, then the sales reps will sell hardware units — a transactional sale pushing low-margin hardware devices.

Conversely, if your objective is to sell more MPS "bundled" solutions that will generate greater and more profitable revenue, then compensating the sales reps on the revenue annuity that results from a MPS sale and paying them on a more lucrative incentive structure (while recognizing the longer sale cycle) will motivate sales reps to NOT just sell boxes, but to sell bundled MPS solutions. The result: Your company will make more money selling MPS and you will increase the odds of retaining the MPS sales reps.

The decision of *how* to compensate sales reps for selling MPS should be made after considering several factors including :

- Your purpose for offering MPS

  For instance, if your goal is to grow revenue by selling a higher percentage of MPS deals than other Imaging & Output deals, then you should consider compensating your reps in a way that motivates them to focus on selling MPS (where they will get paid more in commissions and bonuses and where your margins are higher) rather than just selling low-margin boxes.

- The level of complexity of the compensation structure. For instance, will there be 10 measures-per-sales goals for the reps to chase or will there only be 2 or 3. Simpler is better.

- The back-end process for managing different comp plans. Many dealers I've worked with use dealer automation software such as Digital Gateway's e-Automate, Compass, or some other tool for managing and planning compensation.

Some Possible Approaches to Compensation

Following is a sampling of the approaches some companies in the channel employ to compensate their reps for selling MPS deals.

- *Per-page basis*: pay the sales reps something like 2/10 or 3/10 of a cent for each page that is billed as part of a MPS contract.
- *A percentage of the annual revenue* generated by an account (or a percentage of the incremental growth or profit).
- *A percentage of the monthly recurring fees* that result from a MPS deal.
- *A percent of the quarterly MPS billing* to an account, such as 1/10 of a cent.
- *A percentage of the Total Contract Value* of a MPS deal to be paid up-front, with or without a percentage of the recurring revenue
- *A fixed dollar amount* (say $100) for each printer that is sold as part of a MPS contract

Companies may also want to consider design modifications to their comp plans, measures that relieve some of the pressure on month-to-month quota performance and measures that rewards traditional (box) sellers for prospecting for MPS opportunities. And in the end if all of these measures *still* don't motivate your sales reps to change their behavior, then *mandate* it!

5. Implement a first-rate Assessment methodology

I have a saying that he/she who does the assessment wins the deal. Is this a sure bet? No. But I would estimate that 90% of the companies that conduct the assessment as part of a competitive MPS deal do win the business. Why? Because companies invest time, resources and money into working with a solution provider as the solution provider plans for and conducts the assessment. In such cases, the company relies on the assessment provider as the subject matter expert and, in the end, it may just be easier to dance-wit-what-brung-ya than to engage with another party so far into the MPS investigation process.

The Assessment is where you establish your credibility and expertise with the customer, which makes it understandable why most MPS deals are won or lost at the Assessment stage. So it only makes sense to invest in a strong assessment methodology (whether simple or complex), staff and tools (including a TCO tool like the *Water MPS Assessment TCO Tool*, a pricing tool, etc.) that will help your company become a credible player in the MPS arena.

6. Establish a commitment to offering Managed Print Services

If your company is serious about becoming a MPS solution provider and competing for lucrative MPS deals, then the company must be willing to make the investment in establishing its MPS practice. Far too many companies go into it half-heartedly and wonder why customers don't take them seriously as a MPS player and why they can't win any deals.

Committing to becoming a serious MPS solution provider involves gaining the support of the company's executive management, investing in people, training, processes and tools, investing time and patience (because there will be a learning curve and you will stumble early on) and running that part of the business as though you *have no option* but to make the MPS program successful. Anything short of this simply means that there is no commitment to becoming a serious MPS solution provider and the chances are good that you will fail at it.

TWELVE

Best Practices for Managing and Controlling Personal Printers in the General Office

A Common Challenge

Over the years, I have worked with hundreds of customers on imaging & output initiatives (involving printers & printing, copiers/MFPs, fax and scan), including Managed Print Services, fleet optimization initiatives, fleet refresh projects and imaging & output strategy development to name a few. Regardless of the specific imaging & output project being implemented, one common challenge is consistently faced by I/T managers: What to do about personal printers?

When companies assess their printer fleets they typically find that the number of personal printers (non-shared printers that sit at arms-reach on a user's desk or in a user's office) throughout the environment far exceed the quantity they deem reasonable. My analysis shows that, on average, 46% of the printers in a typical company are personal printers. And in certain industries — such as the Life Sciences where the percentage of personal printers through the company is 90% on average — the percentages are even higher.

Why Personal Printers are a Concern for Many I/T Managers

So, why does having half (or more) of your printer fleet as personal printers disquiet I/T managers? Simply: Cost and support. It is not uncommon for users to acquire personal printers at the local electronics store for $150 and then have them rolled under the company's printer support agreement — oftentimes, even when the printers are still under warranty — at a cost of $180 annually. I once performed a consulting engagement with a company that was overrun by personal printers; nearly everybody had one. The company placed all of these personal printers under a support agreement to the tune of $19-per-month each, plus parts. In effect, they were purchasing two new inkjet printer every year for the employees who had them. In the opinion of most managers, this is wasteful spending.

Increased helpdesk support costs are another reason managers try to stem the growth of personal printers throughout the environment. In most companies when the personal printers have problems, users will place help desk calls and even open trouble tickets for the personal printers! This consumes I/T resources that could be better spent handling "real" problems. And given the lax printer-consumables purchasing controls at some companies, it is also easy for users to place a call to the purchasing department or go on-line to order ink cartridges for their personal printers.

In addition, users will also purchase ink and toner from the local office supplies store and expense the purchase to the company.

Why companies have so many personal printers

When it comes to Imaging & Output devices, users want enough printers available so that they can create their output without having to wait to retrieve it. Users also want the features (e.g. color, speed), functions (e.g. scan-to, fax, private print), performance (e.g. reliability) and proximity (i.e. a short distance from their work station) to make their use of these devices productive. (See: *"Self-Seeking Equilibrium." Why Imaging & Printing Infrastructures "Seek Their Own Level,"* and *the Implications for Profitability*, Tab Edwards 2/08).

And when it comes to *personal* printers, users generally agree on why they are justified in their possession of them: 90% of users surveyed said they "need" personal printers because they print confidential and/or secure documents.

In my experience, the majority of personal printers in corporate offices have come directly from end users, primarily when user departments have the autonomy to purchase equipment as long as the cost is below some specified dollar amount. In most of these cases, the company sets departmental budgets which allow users in the department to acquire office supplies or other "things" they need — without I/T approval — as long as that item is priced within the budget limit. This budget threshold is often in the range of $1,000 to $1,500.

The price of printers is at such a level that users can get really robust printers for under $1,000. But in interviews with some users, they usually opt for the lowest-priced "nice" printer they see at the local office supplies or technology retail store. That device is most often a sub-$200 inkjet printer.

Oftentimes users get fed-up with the print resources available to them in the office so they go out and purchase the inkjet printers *out of their own pockets*. They sometimes even hide the purchase in an expense account. The point to be made is that users will find a way to get the print resource they need if they feel as though their print requirements are not being met by the current printer fleet available to them.

Personal (non-shared) printers are not all bad

Personal printers are not all bad and they do have a role in the workplace. I recommend using personal printers primarily as a niche solution. If there is a function that is not being provided in the shared-printer fleet and there are no plans to provide it, then that could be a good use for a personal printer, assuming the person's work "justifies" it or they can justify ownership based on some valid set of criteria. For instance, a person's physical limitations or one's status in the company could justify the need for a personal printer. I believe personal printers should be provided based on whether or not their need is justifiable according to some set of criteria determined by the company (I have provided a few best practices below).

Best Practices:
Ideas for Improvement

I frequently get requests from customers to offer suggestions on how to deal with the growth of personal, non-shared printers in the general office. Over the years I have worked with customers, large-and-small, that have tried different approaches for managing personal printer-growth and reducing their associated costs. Some of these ideas have worked to varying degrees of success, while many have not. But regardless of the degrees of success of the various strategies, I have compiled the findings of the various strategies and present the results below.

So. What are companies doing to address the issue of superfluous personal printers in the general office? Some of the approaches for dealing with personal printers that I have seen work to varying degrees of success are provided below. It seems as though the stricter the approach, the more effective it is at keeping personal printers under control.

Best Practice #1:

Mandate the removal of personal printers and ban them from the company

This approach is the strictest (and harshest) of all of the best practices. Quite simply, the company's CXO will issue a mandate stating: from this day forward no one can own a personal printer and the personal printers that are in the company today will be removed from your desks. There will be penalties for violating this policy.

The companies that have issued this corporate-mandate approach have done so as part of a serious corporate-wide cost-cutting goal. I worked with a couple of companies that used this approach and, although the most extreme, it was the approach that was *most effective* at nearly eliminating personal printers.

Best Practice #2:
Ownership based on justifiable "need"

Companies using this approach have mandated the removal of all personal printers but have not *banned* them. For users who could justify the need for a personal printer (based on some set of criteria), they would get one and it would be supported centrally. The personal printers in this case would not typically be small inkjet printers, but instead, they would be low-level networkable printers.

The justification criteria commonly used by these companies included the items listed below. If the user could answer "yes" to any one of the criteria, that user would be allowed to keep their printer (if they have one and if it is an approved printer model) or acquire one through the I/T department otherwise.

The criteria include:

1. Are you a company Executive?
2. Are you an Administrative Assistant to an Executive?
3. Do you have a physical limitation that warrants your own printer?
4. For your job, do you need a feature that is not currently available in your department's shared printer fleet? Explain.
5. Does the performance of your job require that you print more than n pages-per-month?
6. Does your job require that you not leave your desk/office frequently (e.g. call center, physician)?
7. Other

I believe this is the best approach because the mandate will eliminate most personal printers, and the users who could benefit from having a personal printer can do so; personal printers do have a place in the office.

Best Practice #3:
Keep the printer until it dies

With this approach, the company restricts the purchase of new personal printers and the personal printers that users currently own will not be supported and supplies will not be provided for them. When the printer dies, the users cannot replace it or have it repaired. The belief is that, over time, as more printers die and are removed from the environment, the company will have removed all personal printers.

This approach is tolerable for users because it gives users an opportunity to wean themselves off the personal printer as opposed to going cold-turkey. However, because users are permitted to keep their personal printers until the printers die, it is easy to bring (sneak) a new printer in and start using it, without detection.

Best Practice #4:
Provide the feature that prompted the purchase of the personal printer

In my years of experience, 95% of users say the reason why they "need" a personal printer is because of confidential printing, personal printing, or because they support a lot of people. When users state a reason that could be satisfied by a *shared* printer (the assumption is that if there is a nearby shared printer that offers the features that prompted the personal printer need), the user will be more willing to sacrifice their personal printer. This is the most popular, but least effective approach because — studies show — that most users will not give-up their personal printer unless they are ordered to do so.

TIME AND MONEY: PERSONAL VS. SHARED PRINTERS

Shared printing can save time *and* money over personal printing!

A concern that managers raise when considering the removal of personal printers is the inconvenience due to the "increase" in the amount of time it would take to walk to the shared printer to retrieve a print job versus simply printing it on a local personal printer. Some managers argue that "expensive" talent waste more (soft cost) dollars when they have to spend more time walking to a shared printer to retrieve their print jobs. However, my analysis shows that it can actually take less time to retrieve a print job from a shared printer located 15-feet away than it does to retrieve that same job from a personal printer in your office. The findings are summarized below. (*This is actual data from an assessment I conducted for a Top-5 U.S.-based global Pharmaceutical manufacturer*).

Parameters:

- Cost of an 8-page print job = $0.40
- Document Production Cost = ($0.83/minute) x (time to retrieve the document)
- Average user salary and benefits = $50/hour

Parameter	35 PPM Shared Printer Located 15' Away	10 PPM Personal Printer on the Desk
Time to print an 8-page document	14-seconds	48-seconds
Round-trip time to retrieve the print job from the printer	19-seconds	0-seconds
Total document production time	33-seconds	48-seconds
Time savings over personal printing	**15-seconds/job**	N/A
Document Production Cost for the 8-page document	**$0.95**	$1.07

THIRTEEN

Insider Trading

The good thing about having a degree of autonomy when writing a book is that you can do things that you could never do when writing a school paper. For example, with this book, there were a couple of things I wanted to share with the reader that just didn't really fit anywhere. So, my solution is to create a chapter called "Insider Trading" and put them all here. It's kinda like the utility drawer in your kitchen: when you're not sure where something goes and you don't feel like figuring it out, it goes in the drawer! And why did I call this chapter "Insider Trading"? Because I am going to share with you some of the dirty little secrets about things that take place when sellers and consultants try to sell you on their Managed print Services-related offerings. I'm sure I'm gonna start receiving all kinds of hate e-mail asking me why I'm telling companies about these practices. The reason is because I want to give the reader all sorts of juicy nuggets of information that will enable them to make informed decisions about MPS. This is true for both buyers and sellers.

What's interesting is that I don't think that any MPS sellers should be *angry* with me for sharing this harmless information with companies. They should instead be *thanking me* for giving them a heads-up and allowing them to prepare for these questions should their MPS prospect ask about them.

The Multi-Year Cost-Savings Illusion

As a customer, have you ever been presented with a multi-year MPS proposal that shows how you will save large sums of money *every year* of the agreement by moving to a MPS solution? It happens all the time. The seller's pitch goes something like this: Dear Customer, by signing this 3-year MPS agreement with my company, we will save you $1M per year for each of the 3 years, for a total of $3M dollars, resulting in a 50% cost savings! And this is how the sales rep calculates these $1M annual cost savings: They will conduct an assessment which shows that the company can reduce its operating costs (from a TCO perspective) by $1M by implementing a MPS solution. Then, the sales rep will say that since this is a 3-year deal, the company will save that $1M *each year*. Wrong! It doesn't work that way. *Table A* below depicts the sales rep's contention.

Example 1
- The sales rep has proposed a 3-year deal
- An Assessment revealed that the company spends $2M per year on Imaging & Output in their current un-managed, inefficient environment.
- The sales rep's proposed MPS solution will cost $1M, for a savings of 50%
- The entire MPS solution will be implemented within Year

I

Why is this wrong? Because once the customer implements the MPS solution in Year 1, *they will no longer be spending $2M per year on Imaging & Output in Years 2 and 3! They will only be spending the $1M cost of the MPS solution!* What the sales rep is *really* saying is that the *Expected Opportunity Loss* (EOL) of not moving to a MPS solution is $3M over three years. EOL is a concept used to express the amount of money (potentially) lost by not selecting the best solution between two options — the status-quo and MPS. That's a lot different than saying the company will save $1M per year for 3 years which assumes the company will continue to save $1M each year, which is not the case in this example or in similar real-world scenarios.

A more plausible proposition from the sales rep would be to say that the company is spending $2M today, and tomorrow, after they have installed MPS, they will only be spending $1M for the same thing for a total savings of 50% — but only with a TCO dollar savings of $1M. This is more reasonable since Years 2 and 3 will become the norm and will not be comparable to any "better" solution proposal, which means there will not be further "significant" cost savings — save for the small future cost savings that will result from continuous process improvements.

So, if I was advising the customer in this case (and I have had to do this several times) I would re-calculate the proposal by creating a deployment schedule which shows what percentage of the MPS solution will be installed by when. This would correctly reflect the fact that as more of the MPS solution is installed, the less money the company will have to pay for Imaging & Output — up until the point that the MPS solution is fully implemented.

To illustrate this concept, let's assume, as stated above, that the customer is paying $2M per year in the Current State for Imaging & Output. And let's assume that the MPS solution will cost $1M per year. Let's also assume that the solution provider will be able to install 10% of the MPS solution per month and that as 10% of the MPS solution (and its associated cost) is installed per month, 10% of the existing Current State fleet (and its associated cost) will be *de-installed*. Therefore, the entire MPS solution will be installed in 10-months and the entire Current State fleet will be de-installed in that same 10-month period. In this scenario, the MPS solution will deliver a cost *reduction* of **38% ($550,000)** in just 10-months (as depicted in *Table B*). This is the realistic way to approximate the cost reduction for a MPS engagement, not the way it is commonly done as illustrated in *Example 1* above.

Table A

The Environment	Annual TCO Year 1	Annual TCO Year 2	Annual TCO Year 3	TOTALS
The Current State (inefficient)	$2,000,000	$2,000,000	$2,000,000	$6,000,000
The Future State with MPS	$1,000,000	$1,000,000	$1,000,000	$3,000,000
Total Savings	$1,000,000	$1,000,000	$1,000,000	**$3,000,000**
				50%

Table B

The Environment	Month 1	Month 2	Month 3	Month 4	Month 5
Current State	$ 180,000	$ 160,000	$ 140,000	$ 120,000	$ 100,000
With MPS	$ 10,000	$ 20,000	$ 30,000	$ 40,000	$ 50,000
Cumulative Costs (MPS + Current)	$ 190,000	$ 180,000	$ 170,000	$ 160,000	$ 150,000
Savings from Current of $2M	$ 10,000	$ 20,000	$ 30,000	$ 40,000	$ 50,000

	Month 6	Month 7	Month 8	Month 9	Month 10	Cumulative 10 Months
	$ 80,000	$ 60,000	$ 40,000	$ 20,000	$ -	$ 900,000
	$ 60,000	$ 70,000	$ 80,000	$ 90,000	$ 100,000	$ 550,000
	$ 140,000	$ 130,000	$ 120,000	$ 110,000	$ 100,000	$ 1,450,000
	$ 60,000	$ 70,000	$ 80,000	$ 90,000	$ 100,000	$ 550,000
						38%

Beware the Charlatan

It seems as though once every two months I get a phone call from a MPS solution provider seeking assistance with selling a competitive Managed Print Services opportunity. Why do I get these bi-monthly calls? Largely because the solution providers are being challenged in an account where they have been the incumbent Imaging & Output product supplier for many years and, suddenly, their incumbent position is challenged by a formidable MPS seller threatening to steal away their install-base business. In response, the solution providers turn to a MPS "expert" for assistance with staving off the competitive threat and assisting them in the battle for their account's MPS business. It seems that these solution providers have attended some consultant's MPS sales training seminar in the past and believed that the MPS consultants could help them in their competitive battle. After all, these consultants were training people on how to *sell* Managed Print Services, so *surely* they could help close an actual real-world deal. The reality, however, appears to be otherwise.

What these solution providers tell me is that when they try to engage the MPS "experts" to help them sell a deal, the "experts" balk and are ultimately forced to admit that they can't actually help these companies *sell* a real MPS deal — let alone a *competitive* one — they just teach MPS sales training classes and write books on the subject. Huh?! If it wasn't true it would be funny. There are actually MPS consultants out there teaching sales reps how to sell MPS but they can't actually do it *themselves*?! I guess the old saying applies in this case: if you can't *do*, teach.

The reality is that the MPS solution area is rife with opportunity and many of these consultants are trying to cash-in on the momentum — especially in the MPS Sales Training space. A colleague of mine — who is quite knowledgeable about Managed Print Services — attended one such MPS sales training seminar and was surprised at how many "erroneous" things the consultant was teaching about how to sell MPS. It was almost as if the trainer had never actually sold a MPS deal before.

Three Simple Questions

If you are considering enrolling in a MPS sales training seminar or contemplating sending your sales team to one, here are a few suggestions that can help you avoid spending your money on a seminar conducted by one of those *hypothetical* MPS sales trainers. So, before enrolling yourself or your sales team in a Managed Print Services sales training seminar, ask the following questions of the seminar's facilitators:

1. *What is your sales background and the sales backgrounds of the other trainers conducting the seminar?* If they haven't carried a bag for the world's most admired companies, walk away, save your money.

2. If they have, then ask: *Have you ever sold a Managed Print Services deal that includes printers, MFPs/digital copiers, supplies & consumables, consulting, professional services, support, software applications and print servers/appliances all combined?* If the answer is "No," walk away, save your money. If the answer is "Yes," then ask the question: *What is the total contract value of the Managed Print Services deals (as described herein) your MPS training team has closed?* If the answer is less than $500Million, walk away, save your money. If the answer is *greater* than $500Million, then ask the trainers/consultants to please list some of the companies they have actually sold Managed Print Services solutions to. If the list does not include any companies you have

ever heard of, walk away.

3. Finally, ask the question: *When was the last time you actually engaged in the sale of a MPS deal and when was the last time you actually helped* close *one?*

The answers to these questions should give you a clear indication of whether or not the sales trainers and consultants conducting the Managed Print Services seminar are credible. They key is to make sure that you are not spending your money on a training class conducted by someone who is trying to train your sales reps on how to do something they themselves have limited current, real-world experience with.

I'd Like to Ask You a Question

FOURTEEN

The Question

After so many years of writing "serious" books and papers, Tab Edwards has decided to do a "fun" project. He is working on a new book and wants to get your response to a simple question.

Please read this first
After reading "The Question" below, please go to www.TabEdwards.com to provide your answer. There is only one rule: we need to get your top-of-mind answer, so ***please provide the first answers that come to your mind***. WE KNOW that you will be tempted to take an hour or four to think about the "perfect" answer, but that is not what we are looking for. We want you to list the first things that you think of after you read *The Question*.

The Scenario

Imagine that you and a total stranger are the only survivors on a boat that is lost at sea. The boat is sinking and can only carry one passenger. You draw straws to see who stays and who must get off the boat; you lose, therefore you must exit the boat which has arrived at a deserted island — an island that you know nothing about.

The Question

If you could take any three (3) things with you onto that deserted island — *things*, not *people* — what three things would you take? By the way, they must be things that you can *carry*.

Please take a minute to visit **www.TabEdwards.com** to give us your answer. Also, share it with your friends so that we can get their answers to *The Question*, too.

Thank You!

The Oxford Hill Press

About the Author

Tab Edwards is a consultant, author and lecturer with more than 25 years of experience in, consulting, sales, entrepreneurship and business management, and is considered to be one of the foremost authorities in the world on the topic of Managed Print Services and optimizing inefficient office output environments.

Tab has worked with companies on Imaging & Output-related projects — and Managed Print Services specifically — since 1998. He developed a Hewlett-Packard MPS assessment process (which he originally named the Hardcopy Operational Assessment or "HOA") and helped develop HP's Managed Print Services offering. He has worked with some of the world's largest companies on global Imaging & Output-related engagements, including 7 of the top 20 *Fortune 500 Corporations*, as well as scores of SMBs and even some Mom-n-Pops. The initiatives include Managed Print Services consulting, MPS sales, MPS Assessments, DISC workshops, Imaging & Output Business Process engagements and Strategy development.

He has held award-winning positions at some of the world's most admired companies, including the IBM Corporation, General Electric, SMCC and Hewlett-Packard where he spent 13 years in sales, global and senior consulting positions.

He is a Certified Six Sigma Green Belt Practitioner and is a frequent guest speaker at seminars and industry events including AIIM (the Association for Information and Image Management) and ImagePrint.

Tab is the author of four books (three on the topic of Imaging & Output and one book about the profession of selling) including: *Paper Problems, Imaging & Output Strategy, MPS: Managed Print Services* and *"Coffee is for Closers Only!"* His fifth book entitled *The Nub* (a book about problem solving) is scheduled for release in 2011 (The Oxford Hill Press).

He holds a Bachelor's degree in Accounting from the University of Pittsburgh, a MBA degree from The Pennsylvania State University and he is a Ph.D. candidate in the field of Management at Pace University's Lubin School of Business.

INDEX

7 Cs 205
  Cashflow 120, 205
7 Cs of Imaging & Output 205
  Community 78, 205
  Coordination 205
    Copiosity 205
  Course 205
    Cognizance 205
7Cs of Imaging & Output 205
  Chart 80, 205

A

Acquisition Process 205
  Problems 205
    Dis-coordination 205
Advertising Agency 20, 205
Analysis 205
  Current State 63, 205
Assessment 205
  Imaging & Output 205
    MPS 205
Asset Management 52, 205
Autonomy of Users to Purchase Hardware 205
  Purchase 32, 205

B

Balance 205
  Efficiency 205
    Importance 112, 205
Benefits to the Seller, MPS 205
  Profit Margins, 205
    Customer Relationship 205
Billing 68, 205
"Boom Box," Buying the 205
  Waste 205
    Excess Capacity 205
Break fix 50, 205
Break fix Support 47, 205
Build the Business Case 192 205

C

Change Management 205
  IMAC 57, 205
Chief Information Officer 15, 205
Chief Output Officer 93, 205
  COO 35, 205
Cognizance 128, 205
Commercial Print / External Print 74 205
Community 153, 205
Compensation, MPS Sales Reps 205
  MPS 205
    Compensation Models 205
Competitive Advantage 205
  TCO 205
    MPS 205
Competitive strength 29, 205

Conflict 206
　Between I/T and Sourcing 92 206
Consultants 17, 206
Consumables 49, 206
　Maintenance Kit 25, 206
Consumer 131, 206
Content Management 71, 206
　Enterprise 72, 206
Contracts 206
　Agreements 57, 206
Convergence 206
　Chart 37, 206
Coordination 81, 206
Copiosity 103, 206
Cost Data 206
　Total Cost of Ownership 206
　　TCO 63, 206
Customer Service 206
　On-line Service Portal 56, 206

D

Data Collection 60, 206
　Quantitative 60, 206
　Qualitative 60, 206
De-installation 51, 206
Deployment Models Types, 206
　Fleet Upgrade. 206
　　Replacement 145, 206
　　Spartan 135, 206
　　Tailored 148, 206
Deployment Models 133, 206
Deployment Project Management 206
　Project Management 51, 206
Device Creep 133, 206
DISC 206

Diagram 124, 206
Dis-coordination 97, 206
Discovery & Design 50, 206
Disposition 206
　Recycling 51, 206
Distinctions 206
　Transactional Sale 206
　Print Services 206
Document Management 20, 71 206
Document Production Cost 206
　DPC 206
　　Calculation Formula 113, 206
Document-related Information Supply Chain 206
　DISC 22, 206

E

Earth Day 158, 206
Efficiency 186, 206
EPA 158, 206
Equilibrium and Balance 110, 206
　Balance 110, 206
　Elements of Equilibrium 206
　　Capacity 206
　　Features 206
　　Functions 206
Excess Capacity 93, 206
Expected Opportunity Loss (EOL) 218, 206
Experience, MPS Helpful 30, 206

F

Facilities 32, 206
Fleet Administration 51, 206
Fleet Management 48, 52, 206

INDEX

"Follow That Paper!" 131, 206
Fortune 500 93, 206
Fragmentation 206
  Chart 34, 207
Freedom 28, 207
Freewheeling 179. 207

G

Gap Analysis 183, 207
General Office 21, 207
General Office Defined 21, 207
"Giant Pharma" 13, 207
"Green" 207
  Environmentally responsible 29 207
  Green Explained 156, 207
  As a MPS Initiative 156, 207
  The 4 Perspectives of Green 162 207
Green Chemistry 207
  The 12 Principles 159, 207
Green Plan 207
  Development of a Green Plan 164 207

H

Hardware 49, 207
Hewlett-Packard 25, 60, 207
  HP's Possible Definition of MPS 21 207
How to Get Started with MPS 207
  Buyers 169, 207
  Sellers 169, 171, 207

I

Imaging & Output 207
  7 Cs 78 207
  Environments 207
    Home Office 207
  Imaging & Output Environments 207
    Mainframe 97, 207
    Specialty 97, 207
    SOHO 98, 207
    General Office 99, 207
  Imaging & Output Evolution 207
    5 Stages 175, 176 207
  Insider Trading 207
    Multi-year Savings Illusion 207
    Charlatans 216,, 207
  Installation 50, 207
  Introduction 13, 207

K

Knowledge 207
  Total Cost of Ownership 28, 207

M

Managed Print Services 15, 207
  Benefits to the Buyer 27, 207
    Operating Cost Savings 22 207
  Elements 207
    Hardware 24, 207
  MPS 207
    Defined 23, 207
  The MPS Process 58, 207
  Revenue 207
    Shrinking Hardware Margins 40 207
  MPS Spectrum of Scope 207

Basic MPS Offering 47, 207
Management, MPS Fleet 207
  Contract 25, 207
Manager of Imaging & Output 35, 207
  MIO 35, 207
  Convergence 35, 207
Market Influences 207
  Key Drivers for Outsourcing 38, 207
  Outsourcing 38, 208
Marketing & Business Integration (MBI) 131, 208
Monitor 208
  Track 208
  Manage 68, 208
MPS 208
  How to Structure 196, 208
  MPS Solution Elements Table 196, 208
MPS Sellers 208
  Getting Started with MPS 193, 208
MPS Solution 208
  Minimum 25, 208
    Printers 24, 208

O

Optimization 53, 68, 208
Order 181, 208
Output Compliance Officer 35, 208
Oxford Hill Consulting 41, 208

P

Paper Trail 131, 208

Where paper goes when printed 130, 208
  MBI 131, 208
Payment Options 53, 208
  Base + Click 55, 208
  Per-Page 55, 208
  Pre-paid Pages 56, 208
Personal Printers 208
  Best Practices for Managing 206, 208
  Guidelines 206, 208
  Personal Printers versus Shared 214, 208
  Time and Money 214, 208
Pharmaceutical Manufacturer 13, 208
Phone Support 56, 208
Preventative Maintenance 50, 208
Print Server Appliance 25, 208
  Network Printing 25, 208
  Print Traffic 25, 208
Print Service Provider 20, 208
Problem Solving Methodology 184, 208
Production 74, 208
  Central Reprographics 74, 208
  CRD 74, 208
Productivity 28, 208
Professional Services 42, 208
Progress 182, 208
Purchasing 32, 208
Purpose 208
  Objective 61, 208

R

Real Estate Department 32, 208

Relationship between Departments 32, 208
   Efficiency 116, 208
   Equilibrium 116, 208
Reliability 28, 208
Resellers 40, 208
Robust MPS 48, 208
   Site Assessment 48, 208

S

Scope of the Book, General Office 19, 208
   Imaging & Output 208
Security 57, 208
"Self Seeking Equilibrium" 117, 208
Self Seeking Equilibrium 107, 208
Senator Gaylord Nelson 158, 209
Service Management 56, 209
Service & Support 25, 209
   Project Management 25, 209
   Administration 209
Shift 35, 209
   Convergence 35, 209
Shift toward MPS 35, 209
   Gartner Prediction 31, 209
Software, for MPS 209
   Applications 49, 209
   Web Jetadmin 25, 209
   Pharos Blueprint 25, 209
Solution Design 209
   Fleet Design 49, 209
   Future State 63. 209
   Model 63, 209
Stop-Start-Learn-Start-Repeat 194, 209

Channel Partners 194, 209
Strategy 173, 209
   Imaging & Output, 209
   Simplified Model for Strategy 187, 209
Study Group, Representative 61, 209
Supplier Diversity 153, 209
Supplies 25, 49, 209
   Replenishment 52, 209

T

Tab Edwards 15, 209
   About the Author 229, 209
   Imaging & Output Strategy Book 175, 209
   Paper Problems Book, 209
   Hardcopy Operational Assessment 61, 209
TCO 209
   Chart 67, 209
The 4 Perspectives of "Green" 209
   Global Warming 162, 209
   Preservation 162, 209
   Energy 209
The Question 226, 209
Three Forces Driving MPS Growth 209
   Market 38, 209
   Customer 37, 209
Toner Leakage Chart 43, 209
   Chart 43, 209
Topological Map 63, 209
Total Cost of Ownership, 209
   Example 94, 209
   TCO 209
   Analysis 209

TCO Elements 121, 209
Total Print Management 72, 209
Tracking 53, 209
  Usage 53, 209
    Billing 53, 209
Training, MPS 209
  Education 52, 209
Transition Management Plan 53, 209
  Training 209
    Communication 53, 209
Two Perspectives 17, 209

U

Understanding 180, 209
Uptime 28, 209
User Feedback 60, 209
VAR 39-42, 46, 49, 209

V

Vendor Influences 40, 210
  Reseller 39-42, 210

W

"Walking Paper," Wasting Time 127, 210
Workflow 70, 210
Workflow Review 63, 210
  "Walking Paper" 63, 210

Made in the USA
San Bernardino, CA
18 July 2014